J. Olson

Lessons Learned
From
Research

Lessons Learned
From
Research

EDITED BY

JUDITH SOWDER AND BONNIE SCHAPPELLE

NATIONAL COUNCIL OF
TEACHERS OF MATHEMATICS

ISBN 0-87353-526-X

Printed in the United States of America

Dedication

This book is dedicated to the many superb teachers with whom we have had the good fortune to work during the course of our careers. They taught us how important research can be in their developing understanding of what it means to teach mathematics. We also dedicate this book to all researchers who care about sharing their research with teachers; we include in this dedication the authors of these chapters, particularly Robbie Case and Cathy Jacobson, both of whom died while this book was being prepared.

WHAT CAN TEACHERS LEARN FROM RESEARCH?

Judith T. Sowder, San Diego State University

MATHEMATICS education researchers are interested in teaching, learning, assessing learning, and designing curricula in K–12 mathematics. They undertake their research with the ultimate goal of affecting what happens in mathematics classrooms. The results of their research appear in a variety of research articles and book chapters. Yet teachers rarely access original research reports, perhaps because researchers tend to write in a style that is often not teacher-friendly. Few teachers ever open an issue of the National Council of Teachers of Mathematics's (NCTM) publication *Journal for Research in Mathematics Education (JRME)* or, for that matter, any other research journal, unless they are assigned to do so for professional development or for a graduate class. On the basis of the difficulties inherent in reading research, one would guess that researchers do not know how to reach a teacher audience. But to get research published in reputable journals, the writer must follow guidelines that in many ways preclude reaching out to teachers. If articles published in *JRME* and elsewhere were written to be easily accessible to teachers, the articles would probably not pass muster as rigorous research. As far back as 1969, Cronbach and Suppes described research as *disciplined inquiry;* it is "inquiry conducted and reported in such a way that the argument can be painstakingly examined" (p. 15). The drawback of this demand for rigor is, of course, that the requirements of preparing research for painstaking examination by peers often render the subsequent report dense and difficult to read and understand for the very people who can most benefit from the research and implement the findings—teachers.

Ways exist, though, to bring research to a teacher audience. Sometimes bridges are constructed between research reports and reports in which authors communicate directly with teachers. Such bridges include research-based articles published in the NCTM's school journals, for example, the "Research Into Practice" articles in *Teaching Children Mathematics*. Other bridges include such books as the three volumes of *Research Ideas for the Classroom* (Jenson, 1993; Owens, 1993; Wilson, 1993), developed with funding from the National Science Foundation. Such books serve a useful purpose, but they are summaries of research studies on particular topics rather than reports of individual research articles, and they do not have the immediate flavor of reading a report of a single research study. Nor do they lead teachers to want to read original research reports and make sense of their implications for practice. Also, they require huge collaborative efforts and are expensive to produce, so they are not produced often enough to keep pace with what research has to offer.

This book provides another kind of bridge to bring research to you, our teacher audience. The chapters are based on individual research articles published in *JRME* during the years that we edited the journal. (Judith Sowder was editor, and Bonnie Schappelle was assistant editor.) All chapters have been rewritten for a teacher audience. We provide commentary that highlights the value of each chapter for our readers. This value is sometimes not apparent to a reader who, for example, reads only chapters dealing with secondary school mathematics and misses chapters on such topics as motivation. Most of the chapters here are relevant to all teachers, even though the relevance might not at first be apparent from the title or the abstract.

Our purpose, though, goes beyond providing you, a classroom teacher, with rewritten research reports that are relevant to what happens in your classroom. Our ultimate goal is to give you information that will lead you to appreciate what research has to offer you and to feel confident to maneuver your way through original research articles to find and take what is of value to you. This skill will not become outdated. Assured of the validity of carefully reviewed research in an article published in *JRME* (or in any other reputable research journal), you are free to bypass much of the detail intended for a researcher audience and to focus only on the parts of the report that are of personal value. In the first section of this book, we reprint a published article from *JRME* and show how it was condensed and rewritten for teachers. We want to help you sort through the detail that researchers must provide and take from the article the elements of the study that will most help you personally: What here applies to your classroom? To your teaching? To your students? We invite you to read other original research from which these chapters were taken and note how we—the authors and editors—have rewritten the articles to highlight those aspects that we think you will find most relevant to your work as teachers.

How Were These Chapters Selected?

First and foremost, we selected articles that have direct connection with K–12 mathematics teaching. Articles dealing with university-level mathematics or with teacher preparation were not considered appropriate for this book unless some tasks or methods of assessment could be useful with K–12 students. Likewise, articles that were purely theoretical in nature were not considered appropriate for this book. Some articles that could have been included had already been rewritten and published for a teacher audience. And some articles were not included simply because of the need to keep the book to a manageable length.

When reading research, one most naturally seeks out the research results and thinks about the implications of those results for the classroom. For example, the message from the results of a study by Linchevski and Kutsher on mixed-ability grouping, reported at the end of the first section, will resonate with many teachers. But research can be valuable in other ways. In a 1990 NCTM yearbook chapter, Silver described two additional features of research reports that can be useful to teachers: the tasks designed for the research and the constructs and perspectives of the researchers.

The tasks used by researchers are carefully designed, and before being used in a research study, they are usually pilot-tested and modified to assure that they measure exactly what the researchers intend to measure. The types of behaviors that these tasks elicit can provide insight into what is being learned and understood. Many of these tasks can be used by a teacher to elicit classroom discussions of important mathematics or as tools to assess learning and understanding. For example, the card-sort task in the Lloyd and Wilson chapter can be used as is or can be modified to provide a teacher with a way to assess students' understanding of function, or even of other topics if the content on the cards is changed.

The constructs and perspectives of the researchers can be useful in classroom settings also. (A *construct* is a concept that has been defined by the researcher in a manner that allows it to be investigated. For example, a researcher might define *self-esteem* in terms of a particular score on a set of items written by the researcher to in some way measure self-esteem.) In the chapter by Pesek and Kirshner, the constructs of *relational learning* and *instrumental learning* are discussed. These terms were coined by Richard Skemp some 30 years ago and were used to distinguish between types of learning. Many teachers have found them valuable because they provide a useful way of thinking and communicating about students' learning. They are as relevant today as when Skemp introduced them.

Organization of the Book

We considered several ways to group the chapters. One obvious way is to use the scheme described above; one section would be devoted to chapters focused on findings; another to chapters focused on tasks; and a third to chapters that describe constructs, perspectives, and frameworks worth consideration by readers. But these three types of chapters will be apparent within other organizational structures.

A second obvious way to organize the chapters is to group them according to instructional level—primary, intermediate, middle, and secondary. This method has the advantage of pointing you to articles that directly focus on what you teach. A distinct disadvantage of this organization is that it can lead you to conclude that the only research relevant to you is that research undertaken at the level at which you teach, but this conclusion is erroneous.

A third way to organize the chapters is to group them around mathematical themes, such as early number, geometry, algebra, and statistics. But some of the articles do not relate to a particular area of mathematics, and this organizational scheme might tend to marginalize the non-content-oriented chapters. Also, when a mathematical topic is a focus of a study, this fact too is quickly apparent and thus does not need to be used as an organizational strategy.

A fourth option, similar to the third, is to organize the chapters not by mathematical topics but rather by areas of interest, such as motivation, tracking, international comparisons, and the effects of cultural differences. But many of these chapters do not easily fit under these or similar headings.

A fifth method is to group by type of study: Does the study focus on learning? On teaching? On curricula? On assessment? This organizational structure is the best match with our goals, but of course many of the chapters here would fit under more than one of these four umbrellas. Although we recognize that overlaps occur, we have decided to use this structure to group the chapters. The existence of overlaps provides occasions for you to disagree with our placement of chapters and to impose a personal reorganization. We will happily acquiesce, because a personal reorganization indicates that you have read and reflected on the ideas in a particular chapter.

For each section, we provide a short introduction to the research studies included in the section. We discuss the research implications for the classroom in terms of the value of the three elements described earlier: research findings, research methods and tasks, and research constructs and perspectives. We provide information about the instructional level and the mathematical focus of each study, but we also discuss the study's relevance beyond that level and how the ideas in the chapter could be extended to other mathematical content areas.

The chapters are uniformly structured. Each chapter begins with an abstract. The use of an abstract is common in research—it provides the reader with a quick overview of the content of the chapter. Thus, the reader can determine from the abstract whether the chapter contains particular information, research on algebra learning, for example. The abstracts here provide some information on the original articles but focus on what is contained in these adapted chapters. A footnote at the beginning of each chapter states the source of the original article.

Each chapter is truly a report of all or a part of an original study or collection of studies. Some authors have undertaken subsequent research on the same topics that they have written about here, but new information is not included. We wanted each chapter to report on one particular research article published in *JRME*. As a reader, you can refer to the original article for additional detail that you might want. In particular, you might want to know where to find more research on the topic in question, and the references in the original article are more extensive than in the modified chapter. Additional evidence, additional protocols of student interviews, and reports on statistical analyses can also be found in the original articles. Most chapters here contain information on the purpose of a research study, brief descriptions of how the study was carried out, the results, and some discussion of the implications of the results. Other chapters contain sets of tasks that we think readers will find interesting. Some background information is included to provide a context for the tasks, but the tasks can often be modified for other settings. Yet other chapters contain only research perspectives. In these instances, the original articles could have been included as longer chapters, but we wanted to give readers examples of studies in which researchers' perspectives and constructs can be useful apart from the results of the study.

In the first section, we include an original research report from the *Journal for Research in Mathematics Education*, followed by a chapter in which the report has been rewritten for a teacher audience. We then discuss how the chapters differ. Although we provide only one example of an originally published article and the rewritten report, we sincerely hope that some of you will go back to other original reports and compare them with the chapters as modified here. This experience will help you better understand how to read research. And, of course, we ultimately hope that you will gain the confidence to read original research articles and take from them results, tasks, and constructs useful for your own teaching.

Although this book focuses on reading and understanding research reports, teachers can also undertake research projects on their own. We believe that practitioner research can profit from familiarity with what is entailed in undertaking a good project. Reading research can help one develop an awareness of the need for careful sampling, for piloting materials and tests that are developed for the research, for ways of avoiding bias, and for careful interpretation of results.

Finally, we hope that researchers themselves will become more interested in rewriting reports of their research studies for a teacher audience. This rewriting is not always an easy task and is often one that, unfortunately, is not rewarded by the authors' institutions. Both teachers and researchers, however, need to find ways to make certain that research influences school practices.

REFERENCES

Cronbach, L. J., & Suppes, P. (Eds.). (1969). *Research for tomorrow's schools: Disciplined inquiry for education.* New York: Macmillan.

Jenson, J. J. (Ed.). (1993). *Research ideas for the classroom: Early childhood mathematics.* Reston, VA: National Council of Teachers of Mathematics, and New York: Macmillan.

Owens, D. T. (Ed.). (1993). *Research ideas for the classroom: Middle grades mathematics.* Reston, VA: National Council of Teachers of Mathematics, and New York: Macmillan.

Silver, E. A. (1990). Contributions of research to practice: Applying findings, methods, and perspectives. In T. J. Cooney (Ed.), *Teaching and learning mathematics in the 1990s* (pp. 1–11). Reston, VA: National Council of Teachers of Mathematics.

Wilson, P. (Ed.). (1993). *Research ideas for the classroom: High school mathematics.* Reston, VA: National Council of Teachers of Mathematics.

SECTION I

RESEARCH RELATED TO TEACHING: INTRODUCTION

THIS section is devoted to research that is related to teaching. One might wonder, given the titles here, why some chapters were placed in this section. We thought that they addressed primarily apporaches that are or are not effective instruction or introduced factors that teachers could, in reflection, relate to instruction.

Chapter 1. "Findings From Research on Motivation in Mathematics Education: What Matters in Coming to Value Mathematics," by James A. Middleton and Photini A. Spanias

For many years research studies on how to motivate students to learn mathematics were hard to locate. Yet *JRME* has been fortunate enough to have received and published two recent manuscripts on this topic.

The authors of this chapter present a comprehensive review of the existing research on motivation, much of it from outside mathematics education research but relevant to mathematics learning. Five powerful statements about motivation are made: Motivations are learned; motivation hinges on students' interpretations of their successes and failures; intrinsic motivation is better than engagement for a reward; inequities are influenced by how different groups are taught to view mathematics; and, last but not least, teachers matter. Each statement is discussed in terms of the evidence of research that buttresses the arguments made.

The authors clearly believe that the last statement is the most important of the five. The tasks teachers design, the interactive feedback they give their students, and their expectations of their students all make a difference in how their students learn. When students receive coherent and consistent experiences doing challenging mathematics, with appropriate structuring and mediation of tasks, they are likely to recognize and take pride in those times when they have learned something significant.

Because this chapter is a review of research, we include far more references than are included in the other chapters. Sometime teachers find references within the text distracting, but in a review chapter, avoiding multiple references is not feasible or appropriate.

Lessons learned

The statements made here about motivation are important to teachers at all levels. A profitable professional-development day could be wisely spent discussing and reflecting on the authors' five main points and their implications for practice.

Chapter 2. "Motivating Students by Teaching for Understanding," by Elham Kazemi and Deborah Stipek

This chapter is the second in which the focus is on motivation. In the study on which this chapter is based, Kazemi and her colleagues used their review of the research on motivation to guide the data collection and interpretation of the results of a large-scale study of relationships among teaching practices, student motivation, and students' learning of fractions. Teaching practices *did* affect motivation and appeared to support conceptual learning. The results of this study provide support for teachers who help their students focus on the learning goals of understanding and improvement rather than on the performance goals of getting right answers and appearing to do well.

One important outcome of this study was the finding of a close relationship between the recommendations from the literature on motivation and the recommendations from mathematics-reform literature. The mathematics-reform literature promotes a focus on learning and understanding, feeling self-confident as mathematics learners, being willing to take risks and attempt challenging tasks, enjoying mathematical activities, and having positive feelings about mathematics. The authors show how each of these foci is reflected in the literature on achievement motivation.

Lessons learned

Many teachers of mathematics will find this chapter supportive of their attempts to teach mathematics for understanding through the use of challenging tasks that can lead to real student satisfaction and through their expectations that every student can succeed. The lessons learned here closely parallel the lessons taken from the previous chapter.

Chapter 3. "Mathematics Learning in Multiple Environments," by Tracy Noble, Ricardo Nemirovsky, Tracey Wright, and Cornelia Tierney

The learning framework introduced in this chapter was used to study development of students' understanding of the mathematics of change. By itself, this framework presents new ideas about learning that are applicable far beyond the study reported in the original article, which is not included here. The framework focuses on mathematical understanding; the underlying message is that the development of a student's mathematical understanding depends on the mathematical environment in which the student is situated. What are the components of an environment that is conducive to mathematical learning? The chapter could quite easily fit into the section on learning, but we have placed it in this section on teaching because, for the most part, the teacher controls the environment in which students learn.

The authors present a question that is being widely discussed among learning psychologists and educators: Where does the mathematics that is being learned reside? Some teachers seem to believe that mathematics resides in the manipulatives and tools we use in the classroom and that their use guarantees learning. Others argue that mathematics resides not in the materials but rather in the manner in which the materials are used within mathematical environments. The authors claim

that mathematical environments are composed of tools, tasks, expectations, conventions, and rules for using tools and that these environments become "lived-in spaces" in which students develop mathematical understanding.

Lessons learned

Teachers at all levels can reflect on how their own choices and actions affect the mathematical environments of the students they teach and how these choices affect what their students learn.

Chapter 4. "Supporting Students' High-Level Thinking, Reasoning, and Communication in Mathematics," by Marjorie Henningsen and Mary Kay Stein

Mathematical tasks engage students in contexts in which they can think about and do mathematics. The nature of tasks can potentially influence and structure the way that students think about mathematics, and they can limit or broaden students' views of the subject matter with which they are engaged. Good teachers know that good mathematical tasks provide them with an avenue to teaching high-level, sound mathematics, but sometimes good tasks do not lead to the kind of thinking that teachers expect. Why not? These researchers identify the major factors that influence maintaining students' engagement, including whether students have the prior knowledge on which to build, whether the necessary scaffolding exists, whether students have sufficient time to complete the task, whether high-level performance is modeled, and whether the teacher sustains pressure for explanation and understanding of the task. The authors also identify the three factors that most influence failure: the decline into using procedures without connections to meaning, decline into unsystematic exploration, and decline from doing mathematics to no mathematical activity. Inappropriate allocation of time plays a role in all three of these factors. Thus, subtle changes to a task can change the ways that students learn from that task.

Lessons learned

Although this study was undertaken at the middle school level, teachers will recognize that it has implications for teaching at all levels. Choosing good mathematical tasks is not sufficient; we also need to understand how to make those tasks work for us. When tasks are used successfully, we can

Lessons Learned From Research

identify which factors led to the success and perhaps increase the effectiveness of that task with other groups of students by increasing attention to factors that may have played smaller roles. When tasks fail, we can identify, using the information presented here, reasons for the failure. By working together and observing one another teach, we can learn to pinpoint aspects of our teaching that caused tasks to succeed or fail.

Chapter 5. "Advancing Children's Mathematical Thinking," by Judith L. Fraivillig, Lauren A. Murphy, and Karen C. Fuson

These researchers studied the instructional practices of primary school teachers teaching from the *Everyday Mathematics* curriculum. In their work, they discovered an exceptional teacher who was able to engage her students in high-level mathematical discussions. The authors undertook an in-depth analysis of this teacher's practices and developed an instructional framework that they called Advancing Children's Thinking (ACT). The framework is focused on eliciting, supporting, and extending mathematical thinking and on the intersections of these three elements of the framework. Each of the three instructional components has major elements that include behaviors that can indicate whether the components are in place.

Lessons learned

At any level, for teachers whose goal is to engage their students as the teacher in this study did, this framework can provide guidelines for evaluating the effectiveness of their own instruction. A teacher could, for example, videotape a few of her or his classes, then use the framework as a way to reflect on personal pedagogical practices. The framework could also be used in professional development settings. For example, a group of teachers might use the framework as a starting point to develop their own ACT framework and arrange to have their principals or instructional leaders use it as a basis for evaluating their teaching.

This brief chapter provides an excellent example of a construct, in this case of the pedagogical practices that advance student thinking, that has implications at all levels of mathematics instruction.

Chapter 6. "The Empty Number Line in Dutch Second Grade," by Anton S. Klein, Meindert Beishuizen, and Adri Treffers

We teachers sometimes become so attached to the arithmetic procedures that we use and pass on to our students that we are not open to learning other procedures that have been found to produce better student learning. This chapter, by Klein, Beishuizen, and Treffers, challenges our tendency to be close-minded about changing what we teach. In the original *JRME* study, the authors report on a comparison of two programs in the Netherlands, but here the focus is on one procedure, used effectively in both programs, for teaching two-digit addition and subtraction. The simplicity and success of the *empty number line* (meaning that 0 and 1 are not designated) is attractive, and the authors provide convincing arguments for why this procedure works well in their country. This procedure is also used extensively in Great Britain and in Australia, and it should work well in many countries.

Lessons learned

We teachers all need to be reminded that the procedures we teach and sometimes the concrete materials we use are not necessarily the most effective ones for all students. We need to keep open minds about other approaches, while remaining critical until we have been convinced that another approach is better, at least under some circumstances.

Chapters 6 and 7 are described in the Interlude that precedes these chapters.

FINDINGS FROM RESEARCH ON MOTIVATION IN MATHEMATICS EDUCATION: WHAT MATTERS IN COMING TO VALUE MATHEMATICS

James A. Middleton and Photini A. Spanias, Arizona State University

Abstract. Consistencies across research perspectives were found in an examination of research in the area of motivation in mathematics education. The findings in this domain led to a set of generalizable conclusions about the contextual factors, cognitive processes, and benefits of interventions that affect students' and teachers' motivational attitudes.

IN THIS chapter we summarize the last 20 years or so of research on motivation in mathematics and suggest how this research may be useful for classroom teachers struggling with the issue. We address the common dilemmas of how and when to provide incentives for work, how to design tasks that increase the probability that more students will become mentally engaged, how to discern and mediate disabling motivational patterns, and what practices typically engage students to think deeply.

In this review, we can summarize the solid, verifiable evidence on what factors influence motivation in five brief statements:

1. Motivations are learned.
2. Motivation hinges on students' interpretations of their successes and failures.
3. Intrinsic motivation is better than engagement for a reward.
4. Inequities are influenced by how different groups are taught to view mathematics.
5. Teachers matter.

This chapter is adapted from Middleton, J. A., & Spanias, P. A. (1999). Motivation for achievement in mathematics: Findings, generalizations, and criticisms of the research. *Journal for Research in Mathematics Education, 30,* 65–88.

We could cite other, more tentative, research that augments these findings, but if one wanted to say what researchers really "know" about motivation to achieve in mathematics, these five statements would be the cornerstone of that argument. For the remainder of this chapter, we expand these statements, providing evidence and examples that illustrate these findings' practical applications to instruction.

FINDING 1: MOTIVATIONS ARE LEARNED

We note that, in the primary grades, mathematics is one of the most liked subject areas, as reported by children. Students in these grades believe that they are competent and that working hard on problems will result in success. Many first and second graders do not even see ability as something distinctly different from effort (Kloosterman, 1993). From this finding, we can infer that young children view the application of effort as a means of increasing the ability to succeed. Moreover, evidence from studies of children's early mathematical thinking (e.g., Carpenter, 1985) indicates that the vast majority of children are "hard-wired" to learn mathematics. According to this evidence, for a few precious years children enjoy "fiddling" with quantities, use cognitive and material resources at their disposal to help them learn, and work hard to solve mathematical problems.

Because of cultural attitudes about what a good learner is, however, we educators begin to identify which children are "quick" at mathematics and which are "slow" while the children are very young, as early as kindergarten or first grade (Wigfield et al., 1992). Grouping practices that exclusively segregate the faster learners from the slower reinforce these attitudes. By the time they reach the middle grades, students begin to perceive mathematics as a special subject area in which fast students succeed and the rest merely get by or fail (Eccles, Wigfield, & Reuman, 1987). They begin to believe that success is due to innate ability only and that expending effort rarely results in a significant change in their success patterns. "So," students think, "if effort doesn't matter, why work hard?"

Consequently, such attitudes significantly determine mathematics achievement in high school and even in college (Meyer & Fennema, 1985). Students typically do not see mathematics as part of their academic self-concepts, and they try to avoid the anxiety that results from engaging in tasks that continually remind them that they are poor learners (Meece, Wigfield, & Eccles, 1990). What is worse, students tend to attribute their feelings toward mathematics, both good and bad, to the influence of a teacher (Hoyles, 1981). We later address what a teacher can do to alleviate these learned reactions.

FINDING 2: MOTIVATION HINGES ON STUDENTS' INTERPRETATIONS OF THEIR SUCCESSES AND FAILURES

As mentioned previously, the ways in which students view the causes of their successes and failures shape the motivation that they have for current and future engagement in mathematics. Since the early 1970s, most research on motivation in education has focused on this issue. The important finding of this research is that, all else being equal, the effort a person is willing to expend on a task is determined by the expectation that her participation will result in success and is mediated by how much she values the task (Brophy, 1986).

Researchers' best estimates are that a person needs to be successful about 70% of the time for continued engagement to be considered both (a) *challenging* enough to warrant the expenditure of effort and (b) *easy* enough for the student to believe that he or she will experience success (Dickinson & Butt, 1989). Think about it. Being successful 100% of the time is boring. One has no surprises, no novelty, no challenge. Experiencing continual failure is frustrating. Even at chance-level, at which the probability of succeeding is about 50%, the task is not only undesirable but also provides little information as to what the learners are doing *right*, so that they can hone their skills, focus on particularly tricky concepts, and, most important, improve their understandings.

Research shows that when students attribute their successes to ability mediated by effort, they tend to succeed and that when they attribute their failures to lack of ability regardless of effort, they tend to fail. Moreover, when students conceive of ability as *fluid*, or subject to improvement *through* effort, they tend to work even harder in mathematics and, thus, are better achievers than students who believe that ability is *fixed* (Dweck, 1986).

When, because of continual lack of successes and personal attribution of failures to lack of ability (a typical pattern in the school experiences of U.S. students), students think that success in an academic endeavor is unattainable for them, they may experience a debilitating motivational pattern called *learned helplessness*. After continually experiencing failure of this sort, a smart child gives up and tries to fill his or her time with other pursuits that, to the child, seem more worthwhile.

Such student views persist as a result of social and educational environments that (a) place high value on ability and lower value on effort and (b) offer little opportunity for individuals with diverse learning styles to supplement their abilities with sustained effort (Covington, 1984). Because helpless individuals believe that success is virtually unattainable, learned helplessness often becomes a stable behavioral pattern that is highly resistant to change. Moreover, evidence indicates that even the cognitive processing of complex mathematical information is hampered directly by learned helplessness (Dweck, 1986).

One must be careful in interpreting such information, however. For reasons echoed throughout the literature, students also seem to require a healthy appreciation for the role of failure in mathematical problem solving. The likelihood of failure in a task increases with the task's difficulty, thus increasing the value of success (see, e.g., Brophy, 1987). Further, learning appropriate coping strategies for failure is an important dispositional component for developing a strong mathematical self-concept and for building appropriate work-related habits.

Finally, if beliefs about success and failure are influential in developing students' motivational patterns in mathematics, something must be said about how we, as a profession and as a society, define *success* and *failure*. Typical current practice leads students to value speed of computation, correctness of answers, and accuracy in following the teacher's example (Kloosterman, 1993). *Success*, defined this way, results from solving many simple exercises quickly, with emphasis on correct answers and an outside authority as the judge of mathematical power. The mathematical aesthetics of logic, rigor, efficiency, and elegance do not play roles in *success* as defined this way.

FINDING 3: INTRINSIC MOTIVATION IS BETTER THAN ENGAGEMENT FOR A REWARD

An inherent dilemma arises between the conditions of most children's schooling and the factors that we know result in *intrinsic motivation* and better achievement. On the one hand, we want children to be self-disciplined, enthusiastic learners in whatever subject we engage them. On the other hand, children are typically compelled by law, not by choice, to go to school; engaged in academic subjects that they would not have selected; and evaluated through socially visible records of their accomplishments and failures. To stimulate both curiosity *and* perseverance under these conditions is difficult. Given the plethora of forces, both external (e.g., praise, grades, judgments of competence) and internal (e.g., joy, satisfaction, feelings of failure), that interact in the day-to-day motivation of a child, the mathematics teacher must choose from a number of strategies that support deep, significant mathematical thinking.

Making these choices is by no means easy. Sometimes coercing a student into appropriate behavior with treats or other rewards may seem the most expedient route to increasing his time on task or to getting the student to do his work. To these rewards, however, are added "hidden costs" that undermine students' desires to engage in subsequent mathematical activity. Research has shown conclusively that engagement in a task under conditions that make obvious that the activity is merely a means to an end (i.e., merely to receive a treat or to avoid a sanction) diminishes the enjoyment students receive from the task, the depth to which they process the information, and their subsequent tendencies to engage in similar activities in the

future (Lepper & Greene, 1978). Because the presence of the reward is the primary reason for the student to engage, he or she feels less internal, or *intrinsic,* motivation to engage. In the future (i.e., in the real world), when immediate, tangible rewards are absent, the student will be less likely to engage in similar tasks.

The advantages of fostering students' intrinsic motivation include developing higher levels of confidence in their abilities to perform mathematics and greater valuation of, and enjoyment in, mathematical tasks. These positive attitudes support a third benefit: development of positive work habits, such as persistence in the face of failure. Together these features support students' mathematics achievement (Gottfried, 1985; Lehman, 1986; Meece et al., 1990).

The Relationship Between Personal Goals and Intrinsic Motivation

Students tend to have three kinds of personal goals for engaging in any particular behavior pattern. Researchers call the goals in the first category *mastery* or *learning goals*. As the terms indicate, a student who develops goals for learning engages in a task to develop understanding or to bolster skills. Consequently, students with learning goals believe that to be successful in the task, one must work hard, attempt to understand the domain, and collaborate with others who have complementary skills and knowledge (Ames, 1992; Ames & Archer, 1988; Duda & Nicholls, 1992; Dweck, 1986). These goals influence a student to become actively cognitively involved in the academic task. The child enjoys involvement in the activity, wants to learn and develop skills, and consequently applies himself or herself more diligently to achieve the goals.

The second category of goals students have for initiating behavior patterns is termed *performance* or *ego* goals. We prefer the term *ego goals* because use of the word *ego* emphasizes the underlying value of social comparison inherent in such goals. Students who engage in an activity because of ego goals tend to value establishing cognitive or personal superiority over others, and they believe that success *depends* on this kind of social comparison (Duda & Nicholls, 1992). This belief manifests itself in students' comparing grades and vying for attention from the teacher to receive social recognition or, conversely, their avoiding negative judgments of their competence by not speaking up in class. Even when engaged in activities that are intrinsically motivating, students who have ego goals tend to display less persistence in the face of frustration and much less active cognitive engagement than students motivated by learning goals.

A third goal orientation, called *work avoidance,* is particularly distressing from the point of view of a teacher. This orientation is characterized by the belief that working hard is not valued. Students who have work-avoidance goals often focus their behaviors on aspects of the task that are really peripheral to its objectives. For example, a student might believe that success depends primarily on being nice to others in his or her group as opposed to interacting deeply in the mathematical problem solving. Many children who display work avoidance have also developed feelings of learned helplessness in mathematics.

The goal orientations that students create, like their perceptions of the causes of success and failure in mathematics, tend to become general patterns of behavior in the middle grades and remain relatively stable throughout a person's life.

FINDING 4: INEQUITIES ARE INFLUENCED BY HOW DIFFERENT GROUPS ARE TAUGHT TO VIEW MATHEMATICS

An area in which motivational research has stimulated significant change in mathematics education is equity. Women and minorities in general have long been excluded from advanced mathematics, including mathematics that leads to high-paying occupations, such as engineering. A significant portion of these inequities can be explained through a motivational lens.

The research on gender differences in mathematics has shown that females have been socialized to view mathematics as a male domain and to perceive themselves as being less able than males to do mathematics (Fennema & Sherman, 1976). Far more girls than boys believe that their failures are due to a lack of ability in mathematics, lack confidence in their abilities, and believe that success in mathematics is unattainable (Benenson & Dweck, 1986).

Girls, more than boys, tend to identify with their mathematics teachers in developing motivation toward mathematics (Fennema & Peterson, 1984). Unfortunately, the behaviors of their teach-

ers tend to reinforce failure-oriented attributions in girls (Fennema, Peterson, Carpenter, & Lubinski, 1990). Teachers tend to show more concern for boys when they struggle than for girls when they struggle. Teachers call on boys more than on girls during question-and-answer sessions and generally engage in more social interaction with boys than with girls. Teachers tend to believe that success in mathematics is due to high ability more frequently for boys than for girls. Teachers view boys more often than girls as the most successful students in the class. Teachers attribute girls' failure to lack of ability, whereas they more often attribute boys' failure solely to lack of effort (Fennema et al., 1990). All these trends unwittingly undermine girls' achievement motivation by reinforcing failure-oriented attributions and contribute to differential gender-related motivational and achievement patterns in girls versus boys at all ability levels (Jackson & Coutts, 1987).

Recent national assessments (e.g., National Center for Education Statistics, 1996) show that the gender gap in achievement, at least, is closing rapidly. Boys and girls show nearly identical mathematics achievement on the National Assessment of Educational Progress in Grades 4, 8, and 12 on assessments between 1990 and 1996. At the same time, both boys and girls increased in achievement across the board from 1990 to 1996. How this closing gap in achievement is related to motivation is unclear, but we surmise that it is positively related, at the very least, to the trend of girls' taking more advanced mathematics courses in high school. These findings on gender differences are echoed in the research on differences across racial and ethnic groups (e.g., Croom, 1984).

FINDING 5: TEACHERS MATTER

If students realize that their successes are meaningful and result both from their abilities and from high degrees of effort, they are likely to believe that they can do mathematics if they try (Relich, 1984). How, then, can a teacher help a student come to this realization?

Although the task may seem daunting, a number of solid rules-of-thumb gleaned from the literature on improving motivation will serve the mathematics teacher well. In particular, the teacher's engineering the classroom goals to focus on learning goals enables the development of similar goals on an individual level (Cobb et al., 1991, Cobb, Wood,

Yackel, & Perlwitz, 1992). Students in such classrooms are less likely to develop ego goals than are students in more traditional classrooms. Moreover, students in inquiry-oriented classrooms are less inclined to believe that conforming to the teacher's method or "the right way to do it" leads to success in mathematics. Instead, they believe more strongly that the classroom is a place where *success* is defined as "working hard to understand mathematics." Explaining thinking and helping others are natural outgrowths of discourse in such classrooms. Attitudes developed in inquiry-oriented classrooms contribute to increased student performance on conceptual and nonroutine tasks; this performance persists even in the face of poor instruction in future grades (Cobb et al., 1991; Cobb et al., 1992).

Although providing treats and other rewards for schoolwork is generally poor pedagogical practice, on occasion doing so can assist in getting students started on difficult tasks when some initial difficulty in motivating them is experienced. One way that rewards can be helpful and less harmful is when the teacher provides *group* incentives instead of individual incentives. Because the group's achievement is rewarded, children are motivated to help others in the group and are pressured to learn well themselves, thus leading to individual accountability (Slavin, 1984). Helping others then becomes a personal goal, and individual diligence contributes to both group and individual success. This practice helps students attribute their successes to their own contribution, whereas the onus for failures can be distributed across the group.

To facilitate the development of students' intrinsic motivation, teachers must teach knowledge and skills worth learning, not only in terms of the structure of mathematics as a discipline but also in terms of the goals and aspirations of students. Many authors of articles in professional journals advocate the development of interesting contexts within which to situate mathematical activities (e.g., Middleton & Roodhardt, 1997). This appropriate contextualization is not just a "touchy-feely" attempt to appeal to students' interests but has significant motivational and cognitive benefits. The best research available indicates that contextualization is beneficial in at least three ways: (a) It piques students' interest; (b) it stimulates students' imaginations and assists in drawing connections among mathematical and everyday concepts that children hold; and (c) it provides functional mathematical knowledge that is useful in a variety of applications (Bransford et al., 1988; Middleton & Roodhardt, 1997). Newer curricula (e.g., *Investigations in Number, Data, and Space; Mathematics in Context; Connected Mathematics*) have capitalized on these benefits and have built powerful, interesting applications of mathematics as the primary vehicle for developing children's mathematical knowledge.

In addition, our own research indicates that teachers who are more attuned to the motivational predilections of their students are better able to fine tune their classroom practices to facilitate the development of intrinsic motivation in those students (Middleton, 1995). No substitute can be found for knowing your students intimately—knowing what their interests are, what challenges them, and what frustrates them.

CONCLUSIONS

The picture we have painted thus far is a hopeful one. Although in the past some educators have led students to fear and loathe personal engagement in mathematical tasks, one can design curricula and instructional techniques to improve the situation. Motivation to achieve in mathematics is not solely a function of innate mathematical ability. Teachers can affect motivation. Fortunately, student motivation in mathematics is highly influenced by teachers' instructional practices. If appropriate practices are *consistent* over a *long period of time*, children can and do learn to enjoy and value mathematics. Moreover, even if children's histories have been consistently poor over a long period of time, the research reviewed in this chapter indicates that classroom practice can be positively reinvented so that the culture of the classroom can become conducive for students to learn and enjoy mathematics.

Although a nearly overwhelming number of studies address the subject of motivation in mathematics education in some manner, the general picture painted, even across lines of inquiry and views about the nature of mathematics, is quite consistent. This consistency gives us considerable confidence in promoting the set of findings we have reviewed in this chapter. Our reference list is shorter than that of the original study. Even that comprehensive review did not present all the related research available, because we were interested only in those studies that specifically dealt with some form of mathematics, not with studies in language arts or other, even closely related, content. A number of high-quality resources are available to help teachers better understand motivation in general; for example, the book *Motivation in Education: Theory, Research, and Applications* by Pintrich and Schunk (1996) is written in a style that is both intellectually stimulating and practically oriented.

So it seems that hope is to be had after all. Motivation to achieve in mathematics is not solely a product of mathematics ability, nor is it so stable that intervention programs cannot be designed to improve it. Instead, achievement motivation in mathematics is highly influenced by instructional practices, and if appropriate practices are consistent over a long period of time, children can and do learn to enjoy and value mathematics.

REFERENCES

Ames, C. (1992). Classrooms: Goals, structures, and student motivation. *Journal of Educational Psychology, 84*, 261–271.

Ames, C., & Archer, J. (1988). Achievement goals in the classroom: Students' learning strategies and motivation processes. *Journal of Educational Psychology, 80*, 260–267.

Benenson F. F., & Dweck, C. S. (1986). The development of trait explanations and self-evaluations in the academic and social domains. *Child Development, 57*, 1179–1187.

Bransford, J., Hasselbring, T., Barron, B., Kulewicz, S., Littlefield, J., & Goin, L. (1988). Uses of macro-contexts to facilitate mathematical thinking. In R. I. Charles & E. A. Silver (Eds.), *The teaching and assessing of mathematical problem solving* (pp. 125–147). Reston, VA: National Council of Teachers of Mathematics, and Hillsdale, NJ: Erlbaum.

Brophy, J. (1986). Teaching and learning mathematics: Where research should be going. *Journal for Research in Mathematics Education, 17*, 323–346.

Brophy, J. (1987). Socializing students' motivation to learn. In M. L. Maehr & D. A. Kleiber (Eds.), *Advances in motivation and achievement* (Vol. 5, pp. 181–210). Greenwich, CT: JAI Press.

Carpenter, T. P. (1985). How children solve simple word problems. *Education and Urban Society, 17*, 417–425.

Cobb, P., Wood, T., Yackel, E., Nicholls, J., Wheatley, G., Trigatti, B., & Perlwitz, M. (1991). Assessment of a problem-centered second-grade mathematics project. *Journal for Research in Mathematics Education, 22*, 3–29.

Cobb, P., Wood, T., Yackel, E., & Perlwitz, M. (1992). A follow-up assessment of a second-grade problem-centered mathematics project. *Educational Studies in Mathematics, 23*, 483–504.

Covington, M. V. (1984). The self-worth theory of achievement motivation: Findings and implications. *The Elementary School Journal, 85*, 5–20.

Croom, L. (1984). The urban project: A model to help minority students prepare for mathematics-based careers [Brief report]. *Journal for Research in Mathematics Education, 15*, 172–176.

Dickinson, D. J., & Butt, J. A. (1989). The effects of success and failure on high-achieving students. *Education and Treatment of Children, 12*, 243–252.

Duda, J. L., & Nicholls, J. G. (1992). Dimensions of achievement motivation in schoolwork and sport. *Journal of Educational Psychology, 84*, 290–299.

Dweck, C. S. (1986). Motivational processes affecting learning. *American Psychologist, 41*, 1040–1048.

Eccles, J., Wigfield, A., & Reuman, D. (1987, April). *Changes in self-perceptions and values at early adolescence.* Paper presented at the annual meeting of the American Educational Research Association, San Francisco.

Fennema, E., & Peterson, P. L. (1984). *Classroom processes and autonomous learning behavior in mathematics* (Final Report, National Science Foundation, SEB-8109077). Washington, DC: U.S. Government Printing Office.

Fennema, E., Peterson, P. L., Carpenter, T. P., & Lubinski, C. A. (1990). Teachers' attributions and beliefs about girls, boys, and mathematics. *Educational Studies in Mathematics, 21*, 55–69.

Fennema, E., & Sherman, J. (1976, April). *Sex-related differences in mathematics learning: Myths, realities, and related factors.* Paper presented at the American Association for the Advancement of Science, Boston.

Gottfried, A. E. (1985). Academic intrinsic motivation in elementary and junior high school students. *Journal of Educational Psychology, 77,* 631–645.

Hoyles, C. (1981). The pupil's view of mathematics learning. In C. Comiti & G. Vergnaud (Eds.), *Proceedings of the conference of the International Group for the Psychology of Mathematics Education* (Vol. 1, pp. 340–345). Grenoble, France: Authors.

Jackson, L., & Coutts, J. (1987). Measuring behavioral success avoidance in mathematics in dyadic settings. In J. C. Bergeron, N. Herscovics, & C. Kieran (Eds.), *Proceedings of the eleventh annual meeting of the International Group for the Psychology of Mathematics Education* (Vol. I, pp. 84–91). Montreal: Authors.

Kloosterman, P. (1993, April). *Students' views of knowing and learning mathematics: Implications for motivation.* Paper presented at the annual meeting of the American Educational Research Association, Atlanta, GA.

Lehman, C. H. (1986). The adult mathematics learner: Attributions, expectations, achievement. In G. Lappan & R. Even (Eds.), *Proceedings of the eighth annual meeting for the North American Chapter of the International Group for the Psychology of Mathematics Education* (pp. 238–243). East Lansing, MI: Authors.

Lepper, M. R., & Greene, D. (Eds.). (1978). *The hidden costs of reward: New perspectives on the psychology of human motivation.* Hillsdale, NJ: Erlbaum.

Meece, J. L., Wigfield, A., & Eccles, J. S. (1990). Predictors of math anxiety and its influence on young adolescents' course enrollment intentions and performance in mathematics. *Journal of Educational Psychology, 82,* 60–70.

Meyer, M. R., & Fennema, E. (1985). Predicting mathematics achievement for females and males from causal attributions. In S. K. Damarin & M. Shelton (Eds.), *Proceedings of the seventh annual conference of the North American Chapter for the Psychology of Mathematics Education* (pp. 201–206). Columbus, OH: Authors.

Middleton, J. A. (1995). A study of intrinsic motivation in the mathematics classroom: A personal constructs approach. *Journal for Research in Mathematics Education, 26,* 254–279.

Middleton, J. A., & Roodhardt, A. (1997). Using knowledge of story schemas to structure mathematical activity. *Current Issues in Middle Level Education, 6*(1), 40–55.

National Center for Education Statistics. (1996). *NAEP 1996 mathematics report card for the nation and the states: Findings from the National Assessment of Educational Progress.* Washington, DC: Author.

Pintrich, P. R., & Schunk, D. H. (1996). *Motivation in education: Theory, research, and applications.* Englewood Cliffs, NJ: Prentice Hall.

Relich, J. (1984). Learned helplessness in arithmetic: An attributional approach to increased self-efficacy and division skills. In B. Southwell, R. Eyland, M. Cooper, J. Conroy, & K. Collis (Eds.), *Proceedings of the eighth international conference for the Psychology of Mathematics Education* (pp. 487–503). Sydney, Australia: Authors.

Slavin, R. E. (1984). Students motivating students to excel: Cooperative incentives, cooperative tasks, and student achievement. *The Elementary School Journal, 85,* 53–63.

Wigfield, A., Harold, R., Eccles, J., Blumenfeld, P., Aberbach, A., Freedman-Doan, C., & Yoon, K. S. (1992, April). *The structure of children's ability perceptions and achievement values: Age, gender, and domain differences.* Paper presented at the annual meeting of the American Educational Research Association, San Francisco.

CHAPTER 2

MOTIVATING STUDENTS BY TEACHING FOR UNDERSTANDING

Elham Kazemi, University of Washington
Deborah Stipek, Stanford University

Abstract. Achievement-motivation and mathematics-education researchers promote many of the same instructional practices. We assessed connections among instructional practices, motivation, and the learning of fractions. Our results indicated that the practices recommended in both research areas positively affected students' motivation (e.g., focus on learning and understanding, positive emotions, and enjoyment) and conceptual learning related to fractions. We also found that *positive student motivation* was associated with increased skills related to fractions.

A S MATHEMATICS educators, we have created ambitious visions for mathematics classrooms (National Council of Teachers of Mathematics, 1989, 2000). We have imagined classrooms in which students develop fluent, efficient, and sophisticated mathematical ideas. Our publications describe challenging and authentic activities that help students become powerful mathematical thinkers. At the inception of this vision lies an enduring challenge for teachers—how to create classroom environments that motivate their students to be both enthusiastic and deeply engaged in their work. We were curious to see how practices suggested by mathematics education reformers connected with the factors that researchers have found to nurture positive motivation.

This chapter is abstracted from Stipek, D., Salmon, J. M., Givvin, K. B., Kazemi, E., Saxe, G., & Macgyvers, V. L. (1998). The value (and convergence) of practices suggested by motivation research and promoted by mathematics education reformers. *Journal for Research in Mathematics Education, 29,* 465–488.

We first looked at the mathematics-reform and achievement-motivation literatures to see what conclusions about effective practices were drawn in each. Then we examined, in real classrooms, how those practices affected children's learning and motivation. Our research shows that engaging students in mathematics that focuses on developing powerful conceptual understandings in a positive climate—using strategies that mathematics reformers suggest—also increases student self-confidence in, and enthusiasm for, mathematics.

MOTIVATION

We focused on five sets of motivational objectives found in the literature on mathematics reform. According to mathematics education experts, instruction should increase students' (a) *focus on learning and understanding* mathematics concepts as well as on getting right answers, (b) *self-confidence* as mathematics learners, (c) willingness to *take risks* and approach challenging tasks, (d) *enjoyment* in engaging in mathematics activities, and (e) related *positive feelings* (e.g., pride in mastery) about mathematics.

These objectives are also prominent in the achievement-motivation literature. Consider first a focus on learning. Motivation theorists distinguish between *learning goals*, defined as a focus on developing skills, increasing understanding, and achieving mastery, and *performance goals*, a focus on looking smart (e.g., by getting a good grade or public recognition) or avoiding looking incompetent. Mathematics reformers make a similar distinction between students' focus on conceptual versus procedural understanding.

Motivation researchers have found that students who have learning goals are more attentive, select more challenging tasks, persist longer in the face of difficulty, use more effective problem-solving strategies, and learn better, especially at a conceptual level, than do students who have performance goals.

The second motivation objective in the literature on mathematics reform, developing and maintaining students' self-confidence, is a central idea in the achievement-motivation literature. Considerable evidence shows that, in achievement contexts, being confident has important implications for one's behavior—including willingness to approach tasks, exert effort, and take pride in success.

The third objective for students, willingness to take risks, can be seen in students' reactions to having difficulty in regular classroom contexts. We assessed willingness to take risks by measuring students' help-seeking behavior. Motivation studies have shown that many students are fearful of revealing their ignorance or being perceived as "dumb" and that they sometimes perceive asking questions as risky. Thus, some students give up or stick with an ineffective strategy rather than seek help.

Students' enjoyment of mathematics, the fourth motivational objective noted in the mathematics-reform literature, is an essential component of intrinsic motivation, which researchers have shown to have many benefits. When compared with students who do not enjoy tasks, students who enjoy tasks stick with them longer, use active problem-solving strategies, show more intensity and greater creativity, and are more flexible in their thinking.

The fifth student objective, positive emotions (especially pride), is also prominent in most theories of motivation. Achievement-motivation theorists suggest that positive emotions produce a desire to approach, and that negative emotions (except guilt) produce a desire to avoid, achievement tasks (see Stipek, 1999, and Weiner, 1992, for reviews of the motivation concepts).

How do teachers structure their classrooms and instruction to achieve these objectives related to student motivation? The literatures on achieve-ment motivation and mathematics reform show the value of many of the same instructional practices.

INSTRUCTIONAL PRACTICES

To help students develop flexible understandings of how mathematical ideas are connected, mathematics reformers recommend that teachers ask students to seek alternative solutions. They suggest giving students opportunities to engage in mathematical conversations and incorporating students' inadequate solutions into instruction by having students use mathematical arguments to decide which answer is right. Instruction should focus on understanding mathematical concepts and developing fluency with mathematical operations and ideas. Reformers also recommend giving substantive feedback, not just scores or grades, on assignments (e.g., Ball, 1993; Carpenter & Fennema, 1991; Kazemi & Stipek, 2001; Lampert, 1991; Prawat, Remillard, Putnam, & Heaton, 1992).

Achievement-motivation researchers have demonstrated the value of these instructional strategies promoted by reformers of mathematics instruction. For example, those who study students' goal orientations have suggested that learning goals are fostered in settings in which students' attention is focused on understanding and in which mistakes are treated as a natural, helpful part of learning. Incorporating errors into instruction is one strategy for conveying the message that errors are a part of the learning process. Students' learning goals are also nurtured by feedback that is substantive and focuses on understanding and mastery more than by grades that reflect relative performance of students. Research shows that the emphasis on substantive, informative evaluation, discussed in the mathematics-reform literature, should also contribute to intrinsic interest in mathematics tasks.

Instructional approaches that foster learning goals also encourage students to engage willingly in challenging tasks and take risks. Research on help-seeking indicates further that students' concerns about revealing ignorance by asking questions might be overcome in a classroom context in which students focus on learning and understanding and the teacher responds supportively to questions.

Practices that focus students' attention on learning and developing understanding instead of on performing better than others should also contribute substantially to students' feelings of competence. *All* students can expect to succeed if *success* is defined in terms of developing mastery. In contrast, when *success* is defined competitively, failure (and therefore a low perception of competence) is guaranteed for some students.

Reformers in mathematics education promote students' autonomous, active engagement with mathematical ideas and their construction of mathematical concepts. Theory and previous research in the field of motivation show that instructional practices that provide students some autonomy should foster feelings of control and more enjoyment in mathematics tasks than practices in which teachers play more traditional authoritative and directive roles.

Studies on intrinsic motivation have shown that individuals also enjoy and choose to engage in tasks that are moderately difficult, vary in format, and are personally meaningful. The multidimensional, real-life tasks recommended by mathematics reformers therefore also foster greater enjoyment than traditional assignments involving repetitive computations that can be done quickly.

According to the literature on motivation, any practices that enhance feelings of competence should also foster feelings of pride. One might expect, furthermore, that some practices described above, such as incorporating students' inadequate solutions into the lesson, should minimize negative emotions such as embarrassment or humiliation. Teachers' own enthusiasm for mathematics and their emotional tone in their instruction (e.g., positive and supportive versus negative and critical) may affect the emotional experiences of students while they are doing mathematics.

Our review of these two distinct literatures reveals that reform-mathematics educators and motivation researchers recommend similar classroom practices, even though they have different purposes in mind. We studied the connections among teaching practices, students' conceptual learning, and motivation related to mathematics in elementary school classrooms.

THE STUDY

Our purpose was to investigate convergence between instructional practices suggested by researchers on achievement motivation and practices promoted in the literature on reform in mathematics instruction and to assess associations among instructional practices, motivation, and learning of fractions.

The participants were fourth-, fifth-, and sixth-grade children from 24 classrooms across several districts in a large urban area. The students represented diverse ethnic backgrounds, with 46 African Americans, 358 Latinos, 49 Asians, 92 Whites, 26 from other ethnic groups, and 53 of unknown ethnicities. All the schools represented in the survey served predominantly low-income families. More than 25% of the students in 8 of the classrooms were predominantly Spanish speaking. Average class sizes were large (about 32 children).

The motivation-related variables discussed above were assessed at the beginning of the year and again in the winter or spring, after the students had completed a unit on fractions. The students completed surveys (see Table 2.1 for examples of survey items), and we observed students' behavior during regular classroom mathematics activities.

We assessed conceptual understanding and computational skills in fractions at the beginning of the year and again after a unit on fractions was completed. The computational items included 7 addition and subtraction problems (e.g., $3/5 + 1/5 =$ ____), 4 equivalence problems (e.g., $1/6 =$ ____), and 4 missing-value equivalence problems (e.g., $3/4 =$ __$/8$). To solve these problems, students could apply procedures without understanding the mathematical concepts. The 13 conceptual items (see Figure 2.1 for examples) were less routine and required students to go beyond computations. For example, "fair share" problems required the children to represent the fractions visually and to produce the fraction notation for a sharing problem.

By analyzing teachers' videotaped fractions lessons, we assessed how the teachers implemented the instructional practices described above. Although the instruction and the specific tasks varied across classrooms, the mathematical ideas (e.g., part-whole relationships, equivalence, reducing and adding fractions) taught were the same in all. For the classroom lessons, we rated (a) how positive the affective climate of the classroom was; (b) how much emphasis the teacher placed on effort, learning and understanding, and student autonomy and initiative; and (c) how often the teacher made comments comparing students to one another and

Table 2.1

Measures of Students' Beliefs, Values, and Goals

Subscale name	Sample items
Perceived Ability	• How good are you at math/fractions? • How hard is it for you to learn new things in math/fractions?
Learning Goals	• How much do/did you care about really understanding math/fractions? When you want to know how well you are doing in math, which of these are important to you? (• how well I understood the work I was doing; • how hard I worked)
Performance Goals	When you want to know how well you are doing in math/fractions, which of these are important to you? (• getting more right or wrong than other kids • doing my work faster than other kids)
Help-Seeking	What do you do when you are having trouble with your math/fractions work? (• ask my teacher for help • ask another student for help)
Positive Emotions	How did you feel when you were doing math/fractions work? (• interested • proud) • How much did you like doing math/fractions work?
Negative Emotions	How do/did you feel when you do/were doing math/fractions work? (• embarrassed • frustrated • upset • worried)
Enjoyment	I like doing math/fractions—it's interesting (• strongly agree • strongly disagree); • How much do you like doing math/fractions?

Note. Items are separated by bullets (•). Students were asked the questions about mathematics in the presurvey; they were asked the corresponding questions about fractions after they had studied fractions.

emphasized speed instead of understanding in completing work.

WHAT WE FOUND

We studied relationships among teaching practices, students' motivation, and students' fractions learning. All the relationships we assessed in this study are depicted in Figure 2.2. Although the figure includes arrows indicating our assumptions about causal directions, we cannot say whether one factor caused another, only how strongly the factors were related to one another.

Students' Motivation

We found important links among the motivation objectives we studied. Students' perceptions of their competence were consistently and strongly associated with the other four motivation variables. Students who rated their mathematics or fractions competencies relatively high were more

focused on learning and mastery and reported more positive emotions, greater enjoyment, and fewer negative emotions than students who rated their competencies relatively low. Although we cannot say what factors caused these relationships, our findings are consistent with past studies that indicate that students' perceptions of their own abilities are central and relate to their emotional experiences and enjoyment.

Our results also support the value of learning goals (being focused on understanding and improvement) over performance goals (being focused on getting right answers and looking good compared to others). Children who claimed to be more concerned about learning than performance had more positive and fewer negative emotional experiences and enjoyed mathematics more than children with the opposite concern. The more students were concerned about how well they performed, the more they claimed to experience negative (beginning of the year) emotions and the less

1. For each picture below, write a fraction to show what part is gray.

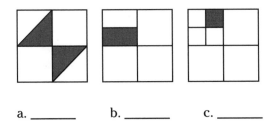

a. _____   b. _____   c. _____

11. Circle all the pictures that show 4/6.

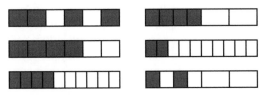

17. <u>Six</u> people are going to share these five chocolate bars equally. Color in **one** person's part.

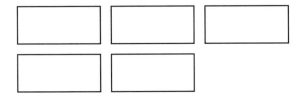

Write a fraction that shows how much **one** person gets. _____

Figure 2.1. Sample of conceptually oriented items from the fractions assessment.

they claimed to experience positive emotions and to enjoy mathematics.

Teacher Practices, Student Motivation, and Learning

This study provides evidence of the effects of teacher practices on student motivation in regular classroom contexts. We measured the affective climate of the classroom by rating teachers' display of enthusiasm and interest in mathematics; their creation of a supportive environment for students to make mistakes and not be put down by classmates; and their display of respect, interest, and sensitivity to their students. The affective climate was the most powerful predictor of student motivation. Children in classrooms with relatively more positive

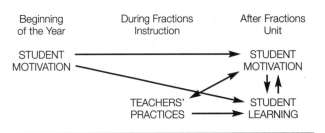

Figure 2.2. Associations tested in the design.

affective climates had stronger learning goals, were more willing to seek help, and experienced more positive emotions than other students did when they were learning fractions.

The central importance of a positive, supportive social climate may be related to other theories and research that show the importance of the relationships students develop with their teachers and classmates. Connell and Wellborn (1991) claimed that *relatedness* (the "need to experience oneself as worthy and capable of love and respect," p. 51) is a basic human need. In their research they showed that students' feelings of relatedness in their classrooms predicted their cognitive, behavioral, and emotional engagement in classroom activities.

Although we did not assess teacher-student relationships directly, students in classrooms with high ratings on the positive-climate dimension likely felt more respected and supported by their teachers than did students in classrooms that were rated low on this dimension. Perhaps students in the former group were able to focus their attention on learning and to ask questions when they were having difficulty, because they did not worry that their teachers' support and positive regard would be withdrawn if they performed poorly or revealed their ignorance.

Students in classrooms in which teachers emphasized effort, learning, and understanding rather than performance and in which autonomy was encouraged also reported experiencing relatively more positive emotions while doing fractions work and enjoying mathematics relatively more than students with teachers who were less supportive. Teachers' providing substantive, constructive feedback on students' papers was associated with positive emotional experiences as well as with a learning orientation. These findings are consistent with intrinsic-motivation theory and with previous research that indicates that feelings of competence (which were likely to have resulted from teachers' emphasis on learning and personal improvement) and feelings of control (which were presumably fostered by the opportunities to work autonomously) engender enjoyment and intrinsic pleasure in activities.

The set of instructional practices that fostered positive motivation also seemed to support students' conceptual learning. Students in classrooms in which the teacher emphasized effort and learning, de-emphasized performance, and encouraged autonomy made substantially greater gains on the fractions-assessment items that required conceptual understanding than did students in classrooms in which teachers did not focus on the task and learning in these ways. The learning-orientation dimension of teacher practices, however, was not associated with achievement gains on the items students could complete by learning a set of procedures without understanding.

These findings are consistent with those in previous studies and show that orienting students toward learning rather than toward performance engenders more active learning strategies (e.g., reviewing material not understood, asking questions while working, planning, setting goals, monitoring comprehension) and less superficial engagement (copying, guessing, skipping questions). The focus on effort and learning in our high-learning-orientation classrooms likely promoted these active problem-solving strategies, which in turn enhanced conceptual learning.

CONCLUSIONS

The results of this study indicate that instructional practices promoted by experts in mathematics reform to increase students' conceptual understanding also achieve important motivational goals. In brief, good instructional practices, those that enhance students' conceptual learning, are also motivating instructional practices. Fortunately, teachers who are working to improve their students' mathematical competencies *and* their motivation to engage in mathematics activities need not choose between the recommendations of researchers working in these two fields.

REFERENCES

Ball, D. (1993). With an eye toward the mathematical horizon: Dilemmas of teaching elementary school mathematics. *Elementary School Journal*, *93*, 373–397.

Carpenter, T., & Fennema, E. (1991). Research and cognitively guided instruction. In E. Fennema, T. P. Carpenter, & S. J. Lamon (Eds.), *Integrating research on teaching and learning mathematics* (pp. 1–16). Albany: State University of New York.

Connell, J., & Wellborn, J. (1991). Competence, autonomy, and relatedness: A motivational analysis of self-system processes. In M. Gunnar & A. Sroufe (Eds.), *Self-processes and development: The Minnesota Symposia on Child Development* (Vol. 23, pp. 43–77). Hillsdale, NJ: Erlbaum.

Kazemi, E., & Stipek, D. (2001). Promoting conceptual thinking in four upper-elementary classrooms. *The Elementary School Journal, 102,* (59–80).

Lampert, M. (1991). Connecting mathematical teaching and learning. In E. Fennema, T. P. Carpenter, & S. J. Lamon (Eds.), *Integrating research on teaching and learning mathematics* (pp. 121–152). Albany: State University of New York.

National Council of Teachers of Mathematics. (1989). *Curriculum and evaluation standards for school mathematics*. Reston, VA: Author.

National Council of Teachers of Mathematics. (2000). *Principles and standards for school mathematics*. Reston, VA: Author.

Prawat, R. S., Remillard, J., Putnam, R. T., & Heaton, R. M. (1992). Teaching mathematics for understanding: Case studies of four fifth-grade teachers. *The Elementary School Journal*, *93*, 145–152.

Stipek, D. (1999). *Motivation to learn: From theory to practice* (3rd ed.). Needham Heights, MA: Allyn & Bacon.

Weiner, B. (1992). *Human motivation: Metaphors, theories, and research*. Newbury Park, CA: Sage.

MATHEMATICS LEARNING IN MULTIPLE ENVIRONMENTS

Tracy Noble, Ricardo Nemirovsky, Tracey Wright, and Cornelia Tierney
Technology Education Research Center

Abstract. As part of an investigation of how students from elementary through high school learn about the mathematics of change in multiple mathematical environments, 5th-grade students were studied while doing mathematics-of-change activities from the *Investigations* curriculum (Russell, Tierney, Mokros, & Economopoulos, 1998) in multiple mathematical environments. This research shows that students, instead of connecting experiences in different environments by recognizing a core mathematical structure common to all environments, make mathematical environments into lived-in spaces for themselves and connect environments through the development of family resemblances across their experiences. This perspective is explored in this chapter.

I N THE mathematics education community, strong support is found for the view that students should encounter mathematical concepts in multiple mathematical environments. In U.S. schools, fraction bars, counters, drawings, and symbols are often used for teaching students about fractions. Tables, graphs, equations, and sometimes even physical devices are used when functions are studied. Yet teachers and researchers have legitimate concerns about confusing students with too unusual or numerous mathematical environments. How can these multiple environments be used to help students to learn mathematics? Where is the mathematics that students are to learn from any given environment, and how do students make connections among different environments?

A mathematically experienced adult may expect students to see the same mathematics in an activity that she herself, on the basis of long experience with the mathematics and the activity, can see. This expectation may lead her to believe that mathematical materials or activities can, in themselves, convey the mathematical ideas that students are expected to learn and that the core mathematical relationships that adults recognize as common to different activities are the only mathematically relevant connections to be made among these activities. However, researchers have found that students often do not make the connections that they are expected to make and sometimes do not even believe that they should find similar results in different environments. We suggest here an alternative view of learning in multiple environments; in this view we take into account students' ways of inhabiting and making sense of each environment. We also articulate a view of the process of making connections among experiences in different mathematical environments; this view allows for diverse and numerous connections among experiences.

WHERE IS THE MATHEMATICS?

Dienes (1964) clearly articulated some ideas, implicit in the work of other researchers, about

This chapter is adapted from Noble, T., Nemirovsky, R., Wright, T., & Tierney, C. (2001). Experiencing change: The mathematics of change in multiple environments. *Journal for Research in Mathematics Education, 32,* 85–108.

The research reported in this article was supported by the SimCalc project, NSF RED-9353507, James Kaput, Jeremy Roschelle, and Ricardo Nemirovsky, co-PIs, and the Mathematics of Change project, NSF MDR-9155746, Ricardo Nemirovsky, PI. All analyses and opinions reported herein are those of the authors and do not necessarily reflect the views of the funding agency.

where the mathematics that children are to learn is located. Dienes used the term *embodiment* to describe a problem situation in which a mathematical structure (such as the real-number system) can be found. Dienes argued that students should be presented with many embodiments of a mathematical structure to "ensure that eventually only the essentially mathematical structure is retained out of all the embodied situations" (p. 41). In this description, the variation of embodiments makes the essential mathematical structure visible, as an invariant. Dienes described the differences among embodiments as the "dressing up," emphasizing that for any embodiment, "the mathematical relevance lies in the number relationships in the problems, not in the 'dressing up'" (p. 35).

Yet Dienes clearly valued more than just mathematical structures as represented in symbolic form. Dienes emphasized the importance of imagery and storytelling in mathematics, stating that "symbol-manipulation in mathematics is all too often meaningless simply because there is no corresponding transformation of images" (1964, p. 105), and he developed a large number of games and stories for making advanced mathematical topics accessible to children.

Dienes's use of the term *embodiment* may contribute to some educators' assumptions that the mathematics that students are to learn can be found in a manipulative material or other tool. A number of researchers have critiqued the idea that concrete manipulatives, physical devices, or computer software can, in themselves, embody mathematical ideas for a student. Meira (1998) criticized the view that specific physical devices can be more or less "transparent" to students in allowing access to mathematical meaning: "The transparency of devices follows from the very process of using them. That is, the transparency of a device emerges anew in every specific context and is created during activity through specific forms of using the device" (p. 138). Using physical materials or computer software in the classroom may provide new opportunities for learning, but the tools themselves do not ensure that students will learn the intended mathematics.

Lesh and his colleagues (1987) have argued that Dienes's work has been misinterpreted by those who believe that Dienes stated that the mathematics resides in physical materials or computer software. They argued that Dienes meant that children

abstract mathematical relations from their own *activities*—from their *use* of materials, not from the materials alone. This rephrased claim may in turn indicate that although the mathematics does not reside in the materials, it can be conveyed to all students through tightly scripted classroom activities involving the materials. However, researchers who have looked carefully at what students do in classroom or interview settings when given a prescribed task have found that the ways students act and make sense of their actions can vary widely from those that are expected by the designer of the activity. This work indicates that mathematical concepts reside not in physical materials, computer software, or prescribed classroom activities but in students' actions and experiences.

In this chapter we use the word *environment* to describe a configuration of tools (such as manipulative materials, a written number table, or computer software), tasks, expectations, conventions, and rules for tool use negotiated in the classroom. We use this term to emphasize that mathematical environments are "'thinking spaces' for working on ideas" (Ball, 1990, p. 7); they do not contain or represent the mathematics to be learned but may provide opportunities for students to develop mathematical ideas. Of course some environments may provide more opportunities than others for students to work with some set of mathematical ideas, but it is up to the learner to inhabit a mathematical environment and to develop the mathematics that she or he will learn through working in the environment.

When a student inhabits a mathematical environment, populating it with her or his actions, intentions, and interactions with others, we say that the student has made the environment a lived-in space for herself or himself. If the purpose of a mathematical activity is not simply to learn the rules and conventions of an environment but instead is to make the environment a lived-in space for oneself, then a diverse range of actions and intentions may be legitimate parts of that mathematical activity.

To understand how mathematical environments can become lived-in spaces for students and how this process relates to students' developing mathematical understandings, we will use an analogy between a dwelling and our understanding of a lived-in space. What is significant about living in a dwelling that has several rooms? Each room has its own uses, and an activity suitable to do in the kitchen is often inappropriate to do in a bedroom, and vice versa. One is never just in the dwelling but is always in this or that room of the dwelling. Our familiarity with the dwelling grows and develops while we act according to our own purposes and expectations in the various rooms of the dwelling. The similarities and differences among these rooms emerge from our experiences, and both similarities and differences are essential to the meaning that the dwelling comes to have for us. We trace an analogy between this imaginary dwelling and the mathematics of change. No ultimate core principle encapsulates all that this mathematics is, in the same way that no single corner of the dwelling tells us everything about the dwelling, in part because experiencing that corner cannot necessarily tell us how we will experience another corner.

This image of mathematical understanding may seem odd to some readers, especially because a mathematical domain is said to have achieved maturity when all its contents can be derived from a small set of axioms. No achievement in the history of mathematics is more celebrated than the formulation of the Euclidean geometry as entirely derived from five axioms. However, we distinguish between the structure of a mathematical system and the nature of mathematical understanding. Although the derivation of a system from a set of axioms can be an important part of learning mathematics, mathematical understanding arises through numerous and varied experiences that allow one to turn mathematical environments into lived-in spaces for oneself. Each environment engages one with some aspects of the mathematics and not with others, not only because of the attributes of any one environment but also because of the intentions, expectations, and actions one brings to each environment.

HOW DO STUDENTS MAKE CONNECTIONS AMONG ENVIRONMENTS?

If students learn mathematics by turning mathematical environments into lived-in spaces for themselves, then how do students connect their experiences in different environments? We cannot explore this question without considering the work of mathematics educators who study students' use of multiple representations. The term *representations* has often been used to describe tools for representing mathematical ideas, especially when these tools are number tables, Cartesian graphs, and equations. Researchers have emphasized the importance of moving among and connecting multi-

ple representations and have also found that differences among representations are important for students' learning.

To further explore the ways that similarities and differences among multiple environments influence students' learning of mathematics, we propose a perspective in which students learn mathematics through the development of "family resemblances" (Wittgenstein, 1958) across lived-in spaces. In his account of the use of words, Wittgenstein argued that words are used and understood not by "applying" definitions but by noticing "family resemblances" (p. 32) among differing uses of a word. Wittgenstein explored the meaning of the word *game* as it is used to describe ball games, word games, card games, and so on. Wittgenstein argued that if you look at all the "proceedings that we call 'games', … you will not see something that is common to *all*, but similarities, relationships, and a whole series of them at that" (p. 31). He described these similarities and relationships as "family resemblances" (p. 32) because they are like the resemblances among members of a family, any two of whom have some characteristic of height, shape, gait, and so on in common but who rarely all share a single characteristic. Thus, according to Wittgenstein, "'Games' form a family" in the sense of having a "complicated network of similarities overlapping and criss-crossing" (p. 32).

Wittgenstein himself related this idea to mathematical concepts, describing the meaning of the word *number* as follows:

Why do we call something a 'number'? Well, perhaps because it has a—direct—relationship with several things that have hitherto been called number; and this can be said to give an indirect relationship to other things we call the same name. And we extend our concept of number as in spinning a thread we twist fibre on fibre. And the strength of the thread does not reside in the fact that some one fibre runs through its whole length, but in the overlapping of many fibres. (1958, p. 32)

Thus, Wittgenstein argued that no unique or ultimate description accounts for the use of a word such as *number*, only a partial and changing likeness developed through a web of contexts. Being a proficient user of the concept of number is being able to move fluently through extended networks of partial similarities and differences.

It is sometimes important for a teacher and her or his students to step back from a diverse set of activities, ask what they have in common, and reflect on the general mathematical principles that describe the activities. However, we argue that these general principles become meaningful and relevant only to the extent that they are rooted in an ongoing background of experiences.

REFERENCES

Ball, D. L. (1990). *Halves, pieces, and twoths: Constructing representational contexts in teaching fractions* (Craft paper 90-2). East Lansing, MI: National Center for Research on Teacher Education. (ERIC Document Reproduction Service No. ED 324 226)

Dienes, Z. P. (1964). *The power of mathematics*. London: Hutchinson Educational.

Lesh, R., Post, T., & Behr, M. (1987). Dienes revisited: Multiple embodiments in computer environments. In I. Wirszup & R. Streit (Eds.), *Developments in school mathematics education around the world* (pp. 647–680). Reston, VA: National Council of Teachers of Mathematics.

Meira, L. (1998). Making sense of instructional devices: The emergence of transparency in mathematical activity. *Journal for Research in Mathematics Education, 29*, 121–142.

Russell, S. J., Tierney, C., Mokros, J., & Economopoulos, K. (1998). *Investigations in number, data, and space*. White Plains, NY: Dale Seymour.

Wittgenstein, L. (1958). *Philosophical investigations* (3rd ed., G. E. M. Anscombe, Trans.). New York: Macmillan.

SUPPORTING STUDENTS' HIGH-LEVEL THINKING, REASONING, AND COMMUNICATION IN MATHEMATICS

Marjorie Henningsen, American University of Beirut, Lebanon
Mary Kay Stein, University of Pittsburgh

Abstract. For students to develop capacities to "do mathematics," classrooms must become environments in which students engage actively in rich, worthwhile mathematical activity. Classroom-based factors can shape students' engagement with mathematical tasks that were set up to encourage high-level mathematical thinking and reasoning. Student engagement is successfully maintained at a high level when many support factors are present. The level of student engagement declines in several ways and for various reasons.

DURING the past decade, much discussion and concern have been focused on limitations in students' conceptual understanding as well as on their thinking, reasoning, and problem-solving skills in mathematics. In response to these concerns, the National Council of Teachers of Mathematics has published proposed reforms of curriculum, evaluation, and teaching practices commonly found in primary and secondary school mathematics classrooms (NCTM, 2000). Two of the underlying goals of these reform efforts are to enhance students' understanding of mathematics and to help them become better mathematical doers and thinkers.

This chapter is adapted from Henningsen, M., & Stein, M. K. (1997). Mathematical tasks and student cognition: Classroom-based factors that support and inhibit high-level mathematical thinking and reasoning. *Journal for Research in Mathematics Education, 28,* 534–549.

Preparation of this paper was supported by a grant from the Ford Foundation (grant no. 890-0572) for the QUASAR project. Any opinions expressed herein are those of the authors and do not necessarily reflect the views of the Ford Foundation.

What are the characteristics of a mathematical doer and thinker? Answers to this question depend on one's view of the nature of mathematics. A view of mathematics that has gained increasing acceptance over the years is one based on a dynamic and exploratory stance toward the discipline rather than a view of mathematics as a static, structured system of facts, procedures, and concepts. This more dynamic notion of mathematical activity has implications for one's ideas about what students need to learn and the kinds of activities in which students and teachers should engage during classroom interactions. Student learning is seen as the process of acquiring a "mathematical disposition" or a "mathematical point of view" (Schoenfeld, 1992) as well as acquiring mathematical knowledge and tools for working with and constructing knowledge. Having a mathematical disposition is characterized by such activities as looking for and exploring patterns to understand mathematical structures and underlying relationships; using available resources effectively and appropriately to formulate and solve problems; making sense of mathematical ideas; thinking and reasoning in flexible ways; conjecturing, generalizing, justifying, and communicating one's mathematical ideas; and deciding whether mathematical results are reasonable.

MATHEMATICAL TASKS

Good Mathematical Instructional Tasks Are Important

Mathematical tasks are central to student learning because "tasks convey messages about what mathematics is and what doing mathematics entails" (NCTM, 1991, p. 24). The tasks with which students engage provide the contexts in which students learn to think about subject matter, and different tasks place differing cognitive demands on students. Thus, the nature of tasks can potentially influence and structure the way students think and can serve to limit or to broaden students' views of the subject matter with which they are engaged. From their actual experiences with mathematics, students develop their sense of what "doing mathematics" means, and students' primary opportunities to experience mathematics as a discipline are situated in the classroom activities in which they engage.

How feasible is consistently and successfully engaging students with high-level tasks involving "doing mathematics" in the classroom? Academic-task researchers (Doyle, 1983) have noted that high-level tasks are often complex and longer in duration than more routine classroom activities; such tasks are thus more susceptible to various factors that could cause students' engagement to decline into less demanding thought processes. Our purpose in this chapter is to identify, examine, and illustrate ways in which classroom-based factors shape students' engagement with high-level mathematical tasks.

What Are the Difficulties Associated With Implementing High-Level Tasks?

Tasks that are set up to engage students in cognitively demanding activities often evolve into less demanding forms of cognitive activity (Doyle, 1983). Engaging in high-level reasoning and problem solving involves more ambiguity and higher levels of personal risk for students than more routine activities. Such engagement can evoke in students a desire for a reduction in task complexity that, in turn, can lead students to pressure teachers to specify further the procedures for completing the task. Classroom-based work on tasks may tend to drift away from a focus on meaning and understanding as the emphasis of the academic work shifts toward accuracy and speed. Another factor underlying unsuccessful task implementation is a lack of alignment between tasks and students' prior knowledge, interests, and motivation. Such mismatches may cause students to fail to engage with the task in ways that will maintain a high level of cognitive activity.

A complex array of factors is involved in orchestrating classroom activity and balancing classroom-management needs with academic demands. These factors can be rooted in the way that classroom norms are set up, in the motivation and learning dispositions of students, and in the general classroom-management practices in which teachers engage. These factors include the manner in which order is established in the classroom, the physical organization of the environment, the amount of time allotted for various activities, the manner in which transition periods between tasks are handled, the establishment of accountability structures, and the ways in which discipline interventions are handled. However, tasks that begin as cognitively demanding do not always decline, and equally important is understanding the ways in which high-level cognitive demands can be *maintained* when the tasks are implemented in the classroom.

What Are Ways of Supporting the Implementation of High-Level Tasks?

Factors that contribute to the decline of high-level demands, when considered in the reverse, can point to ways of maintaining high-level demands. For example, Doyle (1988) has argued that teachers should be especially attentive to the extent to which meaning is emphasized and the extent to which students are explicitly expected to demonstrate understandings of the mathematics underlying the activities in which they are engaged. Such emphasis can be maintained if explicit connections between the mathematical ideas and the activities with which students engage are frequently drawn. Connections with concepts students already know and understand also play an important role in engaging students in high-level thought processes. Some researchers have pointed out that if cognitively demanding tasks *are* appropriate with respect to students' levels and kinds of prior knowledge, students' cognitive processing during task implementation is more likely to remain at a high level than if they are inappropriate (e.g., Bennett & Desforges, 1988). Also, structuring classroom activity so that appropriate amounts of time are devoted to tasks is important.

The mere presence of high-level mathematical tasks in the classroom will not automatically result in student engagement in doing mathematics. Without engaging in such active processes during classroom instruction, students cannot be expected to develop the capacity to think, reason, and problem solve in mathematically appropriate and powerful ways. Clearly, the ambient classroom environment must actively support students' successful engagement in high-level thinking and reasoning.

We investigated classroom factors that either hinder or support student engagement in high-level mathematical thinking and reasoning at the level of doing mathematics. The context for our investigation consisted of mathematics classrooms that were participating in the Quantitative Understanding: Amplifying Student Achievement and Reasoning (QUASAR) project, a national educational-reform project aimed at fostering and studying the development and implementation of enhanced mathematics instructional programs for students attending middle schools in economically disadvantaged communities (Silver & Stein, 1996).

THE CONCEPTUAL FRAMEWORK FOR THIS STUDY

This study is guided by a conceptual framework based on the construct of *mathematical instructional task*. In this framework, shown in Figure 4.1, a *mathematical task* is defined as a classroom activity, the purpose of which is to focus students' attention on a particular mathematical concept, idea, or skill.

In this framework, mathematical tasks are seen as passing through three phases (represented by the rectangular boxes in Figure 4.1): as written by curriculum developers, as set up by the teacher in the classroom, and as enacted by students during the lesson. The framework further specifies two dimensions of mathematical tasks. The first dimension is *task features*, which are aspects of tasks that mathematics educators have identified as important considerations for the development of mathematical understanding, reasoning, and sense making. These features include multiple solution strategies, multiple representations, and mathematical communication. During the set-up phase, these features refer to the extent to which the task as announced by the teacher encourages students to use more than one strategy, to use multiple representations, and to supply explanations and justifications. During the enactment phase, these features refer to the extent to which students use the features.

The second task dimension, *cognitive demands*, refers to the kind of thinking processes entailed in solving the task as announced by the teacher (during the set-up phase) and the thinking processes in which students engage (during the enactment phase). These thinking processes can range from memorization to the use of procedures and algorithms (with or without attention to concepts,

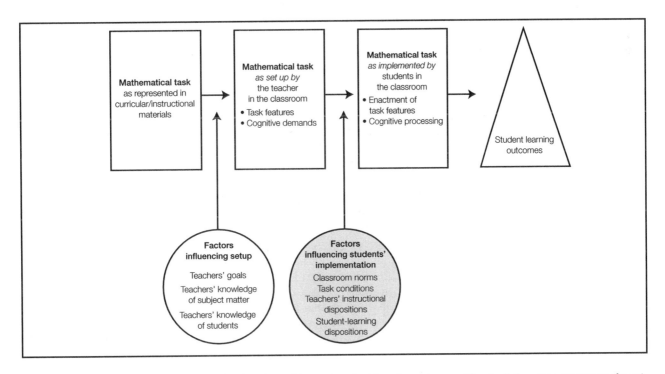

Figure 4.1. Relationships among various task-related variables and student-learning outcomes. The shaded portion represents the primary area under investigation.

understanding, or meaning) to complex thinking and reasoning strategies that would be typical of doing mathematics (e.g., conjecturing, justifying, interpreting). This investigation focuses on this second dimension of cognitive demands and the classroom-based factors that influenced them as tasks proceeded from the set-up to the enactment phase.

According to the framework, the features and cognitive demands of tasks can be transformed between any two successive phases. For example, a task could be set up to require high-level cognitive activity by students, but during the enactment phase it could be transformed by various factors in such a way that student thinking focuses only on procedures with no conceptual connections. The shaded circle in Figure 4.1 represents the classroom-based factors that influence the ways in which student thinking unfolds during the task-enactment phase. These factors include classroom norms, task conditions, and teacher and student dispositions. *Classroom norms* are the established expectations regarding how academic work gets done, by whom, and with what degree of quality and accountability. *Task conditions* are attributes of tasks as they relate to a particular set of students (e.g., the extent to which tasks build on students' prior knowledge, the appropriateness of the amount of time that is provided for students to complete tasks). *Teacher and student dispositions* are relatively enduring features of pedagogical and learning behaviors that tend to influence how teachers and students approach classroom events. Some examples include the extent to which a teacher is willing to let a student struggle with a difficult problem, the kinds of assistance that teachers typically provide to students who are having difficulties, and the extent to which students are willing to persevere in their struggle to solve difficult problems. Through these classroom, task, and teacher and student factors, tasks can be shaped by the ambient classroom culture.

In this chapter, we identify, examine, and illustrate the ways in which classroom-based factors shape students' engagement with high-level mathematical tasks. We identify and describe both those profiles of factors that were associated with maintaining high levels of cognitive demand and those that were associated with each characteristic pattern of decline.

OUR METHOD FOR UNDERTAKING THIS STUDY

Previous to this study, we had gathered extensive data on 144 tasks used in QUASAR classrooms; 58 of these tasks were identified as being set up to encourage doing mathematics. These 58 tasks constitute the data base for the present investigation. During the enactment phases of these 58 tasks, students were observed actively engaging in doing mathematics in 22 of the tasks. In the remaining 36 tasks, students' observed engagement with the task during implementation did *not* exemplify doing mathematics. In 8 tasks, student thinking focused on procedures without connections to underlying meaning; in 11 tasks, students engaged in the form of thinking termed *unsystematic exploration*; and in 10 tasks, student thinking was perceived to have no mathematical focus. In the remaining 7 tasks, students' forms of thinking during the implementation phase represented a variety of categories of cognitive engagement, no one of which was sufficiently well represented to justify its inclusion in our report.

To examine the factors associated with maintenance or decline of the doing-mathematics tasks, we first aggregated and summarized the relevant subset of the factors data from the initial investigation according to the maintenance and decline categories identified above. Within each of these categories, the number of tasks for which each factor was judged to be an influence was calculated. From this information, we constructed frequency graphs so that we could identify factor profiles (i.e., sets of factors judged to be predominant influences in the largest percentage of tasks within each pattern).

WHAT DID WE FIND?

When Are High-Level Cognitive Demands Maintained?

Figure 4.2 shows, for the 22 tasks that remained at the doing-mathematics level during the enactment phase, the percentage of tasks in which each factor was judged to assist students in engaging at that level. The numerals at the top of each bar indicate the number and percentage of tasks for which the particular factor was judged to influence the maintenance of cognitive demands at the level of doing mathematics (e.g., scaffolding was judged to

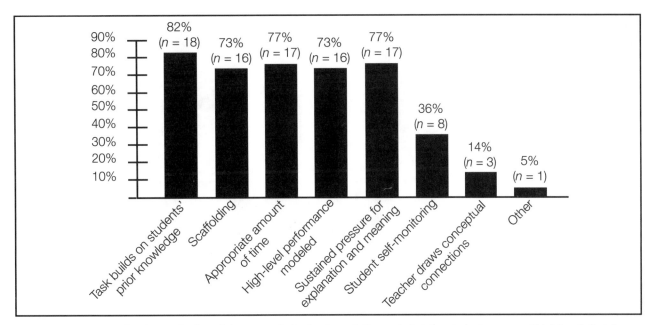

Figure 4.2. Percentage of tasks on which each factor was judged to be an influence in helping students engage at high levels (total number of tasks = 22). Percentages total more than 100 because typically more than one factor was selected for each task.

be influential in 73% of the tasks; scaffolding occurs when a student cannot work through a task on his or her own and a teacher, or more capable peer, provides assistance enabling the student to complete the task alone, without reducing the overall complexity or cognitive demands of the task). Typically, three to five factors per task were believed by the coders to be influential in students' remaining engaged in doing mathematics in particular tasks.

As shown in Figure 4.2, five factors seemed to be primary influences associated with maintaining student engagement at the level of doing mathematics: tasks build on students' prior knowledge (82%), scaffolding (73%), appropriate amount of time (77%), modeling of high-level performance (73%), and sustained press for explanation and meaning (77%). These findings indicate that when tasks successfully maintain student engagement in doing mathematics, these factors would frequently be expected to be in place supporting that high-level engagement in the tasks.

When Do High-Level Cognitive Demands Decline?

The factors that were associated with each of the three types of decline are illustrated in Figure 4.3. In this section, we begin by describing the characteristic factor profiles for each type of decline.

We then look across the three profiles of factors to identify their similarities and differences.

Decline into using procedures without connection to concepts, meaning, and understanding

The factors most frequently judged to influence engagement in those tasks in which students' thinking processes declined into the use of procedures without connection to meaning or understanding were the removal of challenging aspects of the tasks, shifts in focus from understanding to the correctness or completeness of the answer, and inappropriate amounts of time allotted to the tasks. Of these, the factor most often cited was the removal of the challenging aspects of the task during the enactment phase, thus necessitating lower and less sustained levels of thinking, effort, and reasoning by students. Because high-level tasks can be perceived by students (and teachers) as ambiguous or risky, or both, a "pull" is often felt toward reducing their complexity to manage the accompanying anxiety (Doyle, 1988). Reduction in complexity can occur in several ways, including through students' successfully pressuring the teacher to provide explicit procedures for completing the task or the teacher's "taking over" difficult pieces of the task and performing them for the students. When complexity is reduced in these ways, however, the cognitive demands of the task are weakened and students' cognitive processing, in turn, becomes channeled into more predictable and (often) mechanical forms of thinking.

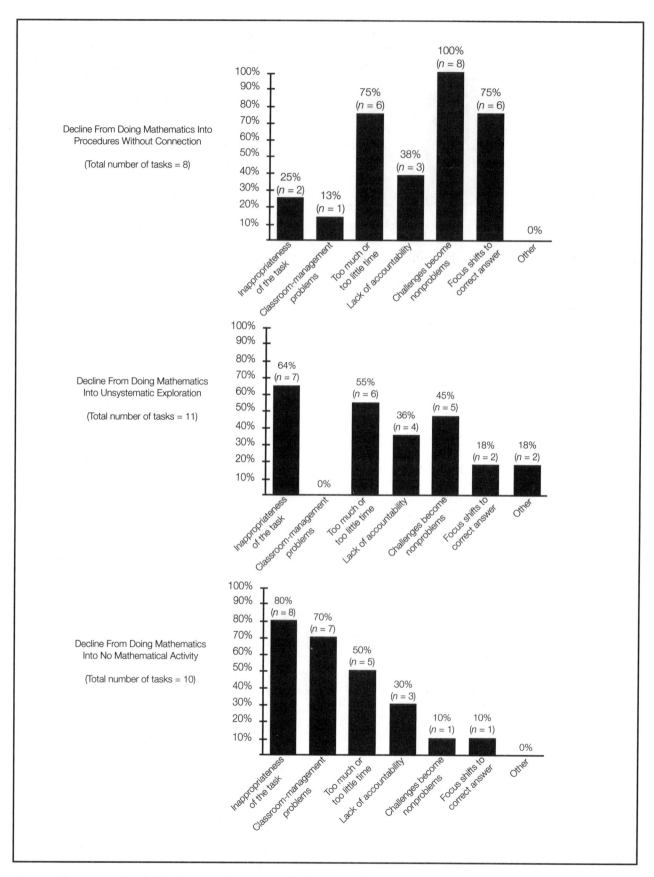

Figure 4.3. Percentage of tasks in which each factor was judged to be an influence in decline of students' engagement at high levels. Percentages total more than 100 because typically more than one factor was selected for each task.

Lessons Learned From Research

Another frequently cited factor was a classroom-based shift in focus away from meaning and understanding toward the completeness or accuracy of answers. The desired outcome of the task becomes defined by the solution rather than by the thinking processes entailed in reaching the solution. Previous mathematical experiences of both teachers and students often lead to such a narrow preoccupation with solutions, at the expense of understanding. This orientation can easily overwhelm tasks that were initially set up to encourage doing mathematics, especially if a focus on process leads to a slowed instructional pace and lack of complete participation by all. Finally, student engagement that declines into procedural forms of thinking often does so because either too much or too little time was devoted to the task. In the examples that we studied, students primarily had too little time to grapple with the important mathematical ideas embedded in the task. A quick pace gives the impression of covering much ground in an efficient manner but often robs students of the time needed to truly engage with the content and to explore and think in ways characteristic of doing mathematics.

Decline into unsystematic exploration

The factors most frequently judged to influence the decline of task engagement into unsatisfying forms of mathematical exploration were inappropriateness of the task for the particular group of students, inappropriate amounts of time allotted for those tasks, and the removal of challenging aspects of the tasks. A task may be inappropriate for various reasons, including low degrees of student motivation, lack of prior knowledge, and lack of suitably specific task expectations. All these reasons relate to the appropriateness of the task for a given group of students. As such, they indicate that an important factor in the success of high-level tasks is the consideration of the relationship between students and task; teachers must know their students well to make intelligent choices regarding the motivational appeal, difficulty level, and degree of task explicitness needed to move students into the right cognitive and affective space so that high-level thinking can occur and progress can be made on the task.

The second most frequently cited factor was inappropriate amounts of time. in contrast with the cause of decline into proceduralized activity, in the majority of tasks for which this factor was judged to be influential in decline into unsystematic explo-

ration, the problem was too much time. When students are not seen to be making headway toward constructing and understanding key ideas, additional time by itself (i.e., without the introduction of additional support factors) appears to exacerbate the situation. Finally, we found that task engagement declined into unsystematic exploration because the challenging aspects of the task were removed. In these cases, however, the removal of the challenge was less often due to the imposition of a procedure and more often due to the subtle alteration of the task so that the main point of the activity was lost or overshadowed.

Another factor that contributed to the decline of task engagement into unsystematic exploration was lack of accountability for high-level products or processes, for example, when students were not expected to justify their methods, when their unclear or incorrect explanations were accepted, or when they were given the impression that their work on these tasks would not "count." In such instances, students circumvent the "real" tasks and tend to focus only on the work for which they will receive a grade.

Decline into no mathematical activity

The factors most frequently judged to influence the decline of tasks into activity with no mathematical substance were inappropriateness of the task, classroom-management problems, and inappropriate amounts of time. Interestingly, classroom-management problems appeared to play a large role in decline to a complete lack of mathematical engagement on the part of the students. This finding shows that teachers were struggling with keeping students under control in addition to keeping them focused on the mathematics (although the two may be subtly interrelated). Once again, inappropriate amounts of time were cited, and, in this instance, the problem was too much time in the majority of tasks in which this factor was cited as influential.

Across the three factor profiles, the decline into procedural thinking appears to be associated with the most clearly discernible, "crispest," pattern of factors. The predominance of the three main factors (contrasted with the relatively weak presence of the other factors) gives a clear picture of activity in classrooms in which such decline occurs. Such sharp distinctions among the predominance of the various factors is not as readily identifiable in the decline into unsystematic exploration. In fact, that profile is the least crisp of all, with several factors contributing moderately to the decline. This finding indicates that the influences that operate in decline into unsystematic exploration are less apparent. Because we did not anticipate declines into unsystematic exploration, the lack of a crisp profile may reflect inadequacies in the ways that our factor categories capture the kinds of classroom scenarios that lead to such decline. The factor profile for decline into no mathematical activity is more sharply differentiated than the profile for decline into unsystematic exploration but is not as crisp as the profile for decline into procedural thinking. The most notable feature of the profile for decline into no mathematical activity is the strong presence of the factor of classroom-management problems.

The one major factor that was shared across the three types of decline was inappropriate amounts of time. Thus, planning for appropriate amounts of time and being flexible about timing decisions when the task-enactment phase unfolds seem to be extremely important to avoid declines of all types. Major factors appearing in two of three profiles of decline were the removal of challenging aspects of the tasks (a factor in declines into proceduralized thinking and unsystematic exploration) and inappropriateness of the task for a particular set of students (a factor in declines into unsystematic exploration and into no mathematical activity). As mentioned earlier, the factor "removal of challenging aspects of tasks" takes on slightly different forms in these two types of decline. The factor of inappropriateness of the task is very broad, covering a variety of reasons for low mathematical engagement; all these reasons relate to the appropriateness of the task for a particular group of students.

SUMMARY AND CONCLUSIONS

Our findings show that a discernible set of factors influences students' task engagement at high levels. Some factors relate to the appropriateness of the task for the students and some to supportive teacher actions such as scaffolding and consistently pressing students to provide meaningful explanations or to make meaningful connections. These findings have implications for the role of the teacher in classrooms in which students are expected to actively engage in doing mathematics. The teacher must not only select and appropriately set up worthwhile mathematical tasks but also proactively and consistently support students' cognitive activity without reducing the complexity and cognitive demands of the task.

Student engagement with tasks declined to lower levels of cognitive activity in different ways and for different reasons. For each of the three patterns of decline, we identified a set of predominant classroom-based factors that contributed to the decline in the cognitive demands of the tasks; however, variation was observed in how sharply distinguishable the factor profiles were across decline patterns. Across all three patterns, one factor, the appropriateness of the amount of time allotted for the task, was a predominant influence, however.

We have continued to explore the classroom-based factors more deeply, examining how student engagement is supported or inhibited during exploratory tasks or investigations that take more than one lesson to complete (Henningsen, 2000). In doing so, we have expanded the list of classroom-based factors and organized them into five categories: task conditions and appropriateness; quality communication of mathematical ideas; sustaining thought over time; tailored assistance; and classroom norms. Using the expanded list of factors, we could describe more completely than was possible in previous research how student thinking can be supported or inhibited.

Since completing this research, we have written and presented our ideas about mathematical tasks and factors that support and inhibit high-level thinking, reasoning, and communication for several audiences: teachers, teacher leaders, professional developers, and school administrators (see Smith & Stein, 1998; Stein & Smith, 1998; Stein, Smith, Henningsen, & Silver, 2000). Without exception, educational practitioners have recognized the patterns of mathematical-task maintenance and decline described in this chapter as well as the factors that shape the ways in which high-level tasks unfold in classrooms. They claim that our work has provided a framework and language useful for talking about classroom events of which they were previously aware but events they were unable to articulate clearly. Currently, this work is represented in several mathematics teacher education courses around the world as well as in our continued work on the development of materials for the professional development of mathematics teacher educators.

REFERENCES

Bennett, N., & Desforges, C. (1988). Matching classroom tasks to students' attainments. *Elementary School Journal, 88*, 221–234.

Doyle, W. (1983). Academic work. *Review of Educational Research, 53*, 159–199.

Doyle, W. (1988). Work in mathematics classes: The context of students' thinking during instruction. *Educational Psychologist, 23*, 167–180.

Henningsen, M. A. (2000). Triumph through adversity: Supporting high-level thinking. *Mathematics Teaching in the Middle School, 6*, 244–248.

National Council of Teachers of Mathematics. (1991). *Professional standards for teaching mathematics*. Reston, VA: Author.

National Council of Teachers of Mathematics. (2000). *Principles and standards for school mathematics*. Reston, VA: Author.

Schoenfeld, A. H. (1992). Learning to think mathematically: Problem solving, metacognition, and sense making in mathematics. In D. A. Grouws (Ed.), *Handbook of research on mathematics teaching and learning* (pp. 334–371). New York: Macmillan.

Silver, E. A., & Stein, M. K. (1995). The QUASAR project: The "revolution of the possible" in mathematics instructional reform in urban middle schools. *Urban Education, 30,* 476–522.

Smith, M. S., & Stein, M. K. (1998). Selecting and creating mathematical tasks: From research to practice. *Mathematics Teaching in the Middle School, 5*, 378–386.

Stein, M. K., & Smith, M. S. (1998). Mathematical tasks as a framework for reflection. *Mathematics Teaching in the Middle School, 3*, 268–275.

Stein, M. K., Smith, M. S., Henningsen, M. A., & Silver, E. A. (2000). *Implementing standards-based mathematics instruction: A casebook for teacher professional development*. New York: Teachers College Press.

CHAPTER 5

ADVANCING CHILDREN'S MATHEMATICAL THINKING

Judith L. Fraivillig, Rider University
Lauren A. Murphy and Karen C. Fuson, Northwestern University

Abstract. A pedagogical framework for Advancing Children's Thinking (ACT) supports children's development of conceptual understanding of mathematics. The framework was synthesized from an in-depth analysis of observed and reported data from a skillful 1st-grade teacher using the *Everyday Mathematics* (EM) curriculum. The ACT framework can contribute to educational research, teacher education, and the design of mathematics curricula.

Abroad-based reform movement focused on teaching and learning in mathematics classrooms and based on curricular and instructional recommendations of the National Council of Teachers of Mathematics (NCTM) *Standards* documents (1989, 1991, 2000) is currently under way in the United States. A major aspect of this movement is the change from traditional classrooms that focus on students' acquiring proficiency in reproducing existing solution methods to classrooms that support helping students construct personally meaningful conceptions of mathematical topics. In another aspect of the movement, researchers examine the instructional changes required to create classrooms that foster children's development of conceptual understanding of mathematics, specifically, the redefined roles and new instructional strategies of classroom teachers.

Establishing classroom norms that support children's development of conceptual understanding of mathematics requires teacher knowledge about both mathematics teaching and children's mathematical thinking. The Cognitively Guided Instruction (CGI) group has identified types of teacher knowledge necessary for creating mathematics classrooms in which student inquiry and explanation of solution methods are encouraged (Carpenter, Fennema, Peterson, Chiang, & Loef, 1989), but to date insufficient articulation has been made of particular instructional strategies that teachers could use once classroom social norms are established. The notion that children construct their own solution methods and mathematical meanings creates confusion about the nature of teachers' roles in inquiry classrooms. Conversations about inquiry-based mathematics classrooms often carry the assumption that teachers should merely elicit children's thinking without intervening. We argue that teachers can and should intervene to advance children's thinking.

In our original study, we investigated the instructional practices of teachers who were using a curriculum that reflects the recommendations for curricular change found in the *Curriculum and Evaluation Standards* (NCTM, 1989). During the course of the study one teacher, Ms. Smith, quickly emerged as a truly exceptional teacher whose mathematics teaching practice was best aligned with *Professional Standards for Teaching Mathematics* (NCTM, 1991). Children in this teacher's class actively engaged in mathematical problem solving,

This chapter is adapted from Fraivillig, J. L., Murphy, L. A., & Fuson, K. C. (1999). Advancing children's mathematical thinking in *Everyday Mathematics* classrooms. *Journal for Research in Mathematics Education, 30,* 148–170. In the original article we described the ACT framework in detail.

This study was supported by the National Science Foundation (NSF) under Grant ESI 9252984. The opinions expressed in this article are those of the authors and do not necessarily reflect the views of the NSF.

eagerly discussed complex mathematical issues, and were excited by intellectual challenges. An in-depth analysis of both this teacher's practice and her reflections about her teaching illuminated the reasons for the qualitative differences in her students' thinking: Many instructional moves made by the teacher were intended to advance children's mathematical thinking.

Our in-depth analysis of aspects of Ms. Smith's mathematics instruction led us to organize the strategies that we observed into a framework for describing effective mathematics teaching. In this chapter we present a framework for examining classrooms and teaching practices.

In our analysis of interview and observational data concerning Ms. Smith's instruction, we found patterns of practices that Ms. Smith used to access children's understanding and to assist and challenge children in their thinking. Three separable, although overlapping, components that composed Ms. Smith's teaching practices were identified: eliciting children's solution methods (eliciting), supporting children's conceptual understanding (supporting), and extending children's mathematical thinking (extending). These three related teaching components and the particular classroom climate in which they occurred form the basis of a framework used to describe observed examples of successful mathematics teaching. This framework, titled the Advancing Children's Thinking (ACT) framework (see the center section of Figure 5.1), assisted us while we continued our research.

However, particular classroom incidents did not always reside neatly within the boundaries of the ACT framework's three instructional components. Some classroom events fell within the intersection(s) between and among components because of their multiple pedagogical functions or their various interpretations. Therefore, particularizing the intersections between and among the three components is helpful (see Figure 5.1). Each of three major components, eliciting, supporting, and extending, is described more fully in Table 5.1.

The ACT framework introduced in this chapter may be useful as a mechanism for organizing complex classroom data and for evaluating instruction. As a common point of reference, the framework can be used by researchers and practitioners to influence and anchor their discourse about mathematical teaching practice. By making explicit the teach-

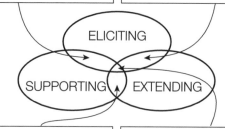

Figure 5.1. Diagram of the Advancing Children's Thinking (ACT) framework and the teaching strategies that reside in the intersections between and among the instructional components.

ing processes associated with a problem-solving orientation in general, the ACT framework can be a useful pedagogical tool for preservice and in-service teacher education. Furthermore, the ACT framework can be used to guide curriculum-design efforts to support teachers' understanding of productive mathematical activity for learners.

REFERENCES

Carpenter, T. P., Fennema, E., Peterson, P. L., Chiang, C. P., & Loef, M. (1989). Using knowledge of children's mathematics thinking in classroom teaching: An experimental study. *American Educational Research Journal, 26,* 499–531.

National Council of Teachers of Mathematics. (1989). *Curriculum and evaluation standards for school mathematics.* Reston, VA: Author.

National Council of Teachers of Mathematics. (1991). *Professional standards for teaching mathematics.* Reston, VA: Author.

National Council of Teachers of Mathematics. (2000). *Principles and standards for school mathematics.* Reston, VA: Author.

Table 5.1

Examples of Instructional Strategies Teachers Employ to Elicit, Support, and Extend Children's Mathematical Thinking

Instructional components of the ACT framework		
Eliciting	Supporting	Extending

Facilitates students' responding

Elicits many solution methods for one problem

Waits for, and listens to, students' descriptions of solution methods

Encourages elaboration of students' responses

Conveys accepting attitude toward students' errors and problem-solving efforts

Promotes collaborative problem solving

Orchestrates classroom discussions

Uses students' explanations for lesson's content

Monitors students' levels of engagement

Decides which students need opportunities to speak publicly or which methods should be discussed

Supports describer's thinking

Reminds students of conceptually similar problem situations

Provides background knowledge

Directs group help for an individual student

Assists individual students in clarifying their own solution methods

Supports listeners' thinking

Provides teacher-led instant replays

Demonstrates teacher-selected solution methods without endorsing the adoption of a particular method

Supports describer's and listeners' thinking

Records symbolic representation of each solution method on the chalkboard

Asks a different student to explain a peer's method

Supports individuals in private help sessions

Encourages students to request assistance (only when needed)

Maintains high standards and expectations for all students

Asks all students to attempt to solve difficult problems and try various solution methods

Encourages mathematical reflection

Encourages students to analyze, compare, and generalize mathematical concepts

Encourages students to consider and discuss interrelationships among concepts

Lists all solution methods on the chalkboard to promote reflection

Goes beyond initial solution methods

Pushes individual students to try alternative solution methods for one problem situation

Promotes use of more efficient solution methods for all students

Uses students' responses, questions, and problems as core lesson

Cultivates love of challenge

CHAPTER 6

THE EMPTY NUMBER LINE IN DUTCH SECOND GRADE

Anton S. Klein, Research Voor Beleid (Research for Policy), Leiden, The Netherlands
Meindert Beishuizen, Leiden University, The Netherlands
Adri Treffers, Utrecht University, The Netherlands

Abstract. Our study was a comparison of 2 experimental programs for teaching mental addition and subtraction in the Dutch 2nd grade. The goal in both programs is to develop greater flexibility in mental arithmetic through use of the empty number line as a new mental model. In this chapter, the programs themselves are not described, but rather the empty number line, which appeared to be a very powerful model for children's learning of addition and subtraction up to 100 in both programs, is introduced and discussed.

A N EVALUATION study in the 1980s in the Netherlands revealed a generally low level of flexibility in children's use of arithmetic strategies. As a result, a new lower grades curriculum was designed; in this curriculum, mental arithmetic played a central role during the first and second grades. Mental arithmetic was seen as a foundation for the further development of flexible computation and problem-solving strategies (Treffers, 1991). Mental calculating could be done not only "in the head" but also by "using one's head": The use of written work was encouraged. Because this view is now dominant in Dutch schools and textbooks, procedures for written (column) arithmetic are not introduced until the third grade.

In our research, we have found empirical evidence that different models have different effects on mental-

computation procedures (Beishuizen, 1993). For example, arithmetic blocks evoked decomposition or place-value strategies for addition and subtraction, whereas the hundred square stimulated a sequential pattern of counting by tens. Arithmetic blocks consist of small cubes that could each represent one unit, a stick of ten units, and a larger cube of 100 units. A hundred square is the set of numbers 1–100 written in 10 rows of 10 columns. Many researchers see place-value and counting-by-ten strategies as basic for addition and subtraction up to 100.

THE EMPTY NUMBER LINE AS A NEW TEACHING MODEL

In the 1980s, researchers in our country found that the hundred square provided a better model for computation than arithmetic blocks, but even the hundred square has been found to be an overly complicated learning aid for weaker pupils. Treffers and De Moor (1990), in their revision of the Dutch primary mathematics curriculum, devised a new format for the old number line, the empty number line up to 100, as a learning aid for addition and subtraction. Children's use of the empty number line is illustrated in Figure 6.1.

We cite several instructional and psychological reasons for using the empty number line as a central model for addition and subtraction.

1. Because of the linear character of the number line, the empty number line is well suited to link up with informal solution procedures.

A structured bead string should be used as an introductory model for the empty number line. This bead string has 100 beads, ordered following a tens

This chapter is adapted from Klein, A. S., Beishuizen, M., & Treffers, A. (1998). The empty number line in Dutch second grades: *Realistic* versus *gradual* program design. *Journal for Research in Mathematics Education, 29,* 443–464.

The research reported in this article was supported by the Dutch National Science Foundation Grant 575-90-607.

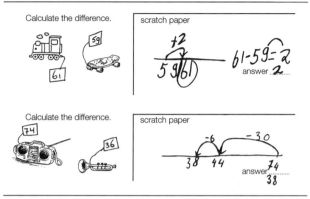

Figure 6.1. Two pupils use the empty number line to solve contextualized difference problems in different ways.

structure: 10 red beads followed by 10 white beads followed by 10 red beads and so on (see Figure 6.2). This structure helps students find a given number and familiarizes them with the positioning of numbers up to 100 and with the quantities that the numbers represent. The tens can serve as a point of reference in two ways: For example, there are 6 tens in 64 and almost 7 tens in 69. After children work with the bead string, the number line can be introduced as a model of the bead string (see Figure 6.2). By using the empty number line, children can extend their counting strategies and raise the sophistication level of their strategies from counting by ones to counting by tens to counting by multiples of 10.

2. The second reason for using the empty number line is that it provides the oppor-

tunity to raise the level of the students' activity.

Not only should a model give students the freedom to develop their own solution procedures, but employing the model should also foster the development of more sophisticated strategies. The empty number line not only allows students to express and communicate their own solution procedures but also facilitates their use of those solution procedures. Marking the steps on the number line as shown in Figure 6.2 functions as a kind of scaffolding by showing which part of the operation has been carried out and which part remains to be done.

3. The empty number line has a natural and transparent character. The format stimulates a mental representation of numbers and number operations (addition and subtraction).

The model seems to be very suitable for the representation and solution of nonstandard context or word problems. The problem representation is clearer and more natural on the empty number line than on the hundred square or arithmetic blocks because the row of numbers is not cut off at each ten. Gravemeijer (1994) reported that for subtraction problems with larger numbers, children were in favor of the adding-on strategy (e.g., Find 75 – 56 by adding on to 56: 4 to make 60, 10 more to make 70, and 5 more to make 75; 4 + 10 + 5 is 19.) Because of its linear character, the empty number line is well suited to making informal solution procedures explicit.

4. Students using the empty number line are cognitively involved in their actions. In contrast, students who use such materials as arithmetic blocks or the hundred square sometimes tend to depend primarily on visualization, which results in a passive "reading off" behavior rather than cognitive involvement in the actions undertaken.

CONCLUSION

In our study, children at the end of Grade 2 were successful on the difficult subtraction problems in the National Arithmetic Test. Pupils' work, as well as incidental classroom observations and teachers' experiences, provided additional clues for the interpretation of its success: During the first half-year in

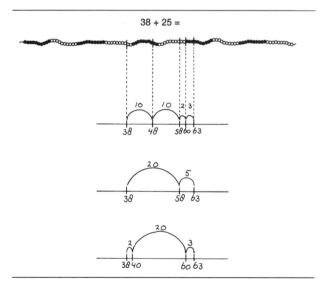

Figure 6.2. The empty number line as a model of the bead string: The problem 38 + 25 is solved in different ways and at different levels.

both programs, the modeling function of the number line supported both procedural operations and problem representation. Evoking children's own mental activity was also a significant function, and this activity became stronger during the second half-year. The success of these students confirmed that the empty number line is a powerful model for instruction.

REFERENCES

Beishuizen, M. (1993). Mental strategies and materials or models for addition and subtraction up to 100 in Dutch second grades. *Journal for Research in Mathematics Education*, *24*, 294–323.

Gravemeijer, K. (1994). Educational development and developmental research in mathematics education. *Journal for Research in Mathematics Education*, *25*, 443–471.

Treffers, A. (1991). Didactical background of a mathematics program for primary education. In L. Streefland (Ed.), *Realistic mathematics education in primary school* (pp. 21–56). Utrecht, The Netherlands: Freudenthal Institute, Utrecht University.

Treffers, A., & De Moor, E. (1990). *Proeve van een nationaal programma van het reken-wiskunde-onderwijs op de basisschool. Deel 2: Basisvaardigheden en cijferen* [Specimen of a national program for primary mathematics teaching. Part 2: Basic mental strategies and written computation]. Tilburg, The Netherlands: Zwijsen.

THIS part of the section contains two chapters. The first is an original article published in the *Journal for Research in Mathematics Education* in November 1998, reprinted exactly as published. The second is that article rewritten for a teacher audience instead of a researcher audience. You are invited to compare the two versions and to study the changes made. We encourage you to make the comparison with a colleague because good collegial discussion may lead to your noticing aspects of the articles that might not otherwise be noted. We hope that doing so will help you read original research reports and find what is of interest to you as a teacher.

In general, an approach to reading a research article might include the following steps:

1. Read the abstract to get a general overview of the report.

2. Even if you do not think that the report itself will be of direct interest to you, at least page through to find the tasks that are used. Are they of interest? Can they be modified for use in your own teaching? For example, could you devise a card sort that is based on the model presented by Lloyd and Wilson in Chapter 28 but that is designed to assess children's understanding of the relationships among fractions, decimals, and percents? Could the tasks used in the Moss and Case study described in Chapter 17 be used with children beyond fourth grade?

3. Also examine the theoretical or conceptual framework used. It may not be of particular interest, but sometimes you will find some gems that make you think about teaching and learning in new ways. For example, the theoretical framework presented by Noble and her colleagues in Chapter 3 gives new ways of

thinking about mathematics-classroom environments.

4. If the abstract leads you to want to know more about the study, then locate, in the first part of the study, the purpose of the study so that you will know how to interpret the results.

5. Most teachers are not interested in the review of all the other research related to the topic, and thus this section of the report can often be skipped. (You will notice that the reports in this book do not contain the literature reviews found in the original articles except in Chapter 1, which is an article focused on reviewing literature on motivation.)

6. Usually in a section called Method, authors tell how they undertook their study. In this section you will find information on the students who participated, the instruments (instructional materials, assessment instruments, tasks) used in the study, and the design of the study. If more detail is provided than you feel you need, skim this section and return to it later if you have questions about the results and how they were obtained. Consider whether the tasks and the instructional and assessment materials are useful, either as presented or modified to suit your purposes.

7. In the Results section, authors provide information on their findings. Once again, they may give more detail than you need, including long verbatim interview protocols or statistical analysis. But you should be able to focus on the findings from the study. For example, after statistically analyzing some data, the authors summarize the results of the statistical tests. You need not understand the statistical tests that

were used. The authors will tell you if the test results were significant and what this significance means.

8. The final section is often called Discussion or Discussion and Implications, although the discussion is sometimes integrated into the Results section. This section may be the most helpful part of the study if you are interested in the findings. The authors discuss what they think the results mean, and often they present implications of the results of the study. You may want to read this part first, then go back to pick up the information you need to understand the discussion.

9. Sometimes researchers undertake a series of closely related studies and report them in the same article as Study 1, Study 2, and so on. Usually, but not always, the final study has the most interesting results, because the previous studies raised questions that are answered in the final study. Sometimes you may be interested in only one of the studies.

10. Share what you learn with your colleagues. Talking about what you learn from a research study always leads to further learning, both for your colleagues and for you.

You might want to test these guidelines by reading Chapter 7 before you read Chapter 8. Take a few minutes to consider what ideas you found of interest in Chapter 7 and how you used the guidelines. Then read the condensed Chapter 8 (based on the article in Chapter 7), which was written for this book. Locate the types of changes made and compare them with the parts of the original on which you focused your attention. Finally, read the Editors' Comments to learn how we, the authors and editors, selected what to keep, what to delete, and what to reword from the original article. You may agree with some of our decisions but disagree with others. If so, you have learned to read research for yourself and no longer need rewritten articles like the ones in this book, in which case we will have succeeded in meeting our goal of helping you learn to profit from research reports.

The remaining sections of the book contain chapters focused on learning, teaching, curricula, and assessment.

TELL ME WITH WHOM YOU'RE LEARNING, AND I'LL TELL YOU HOW MUCH YOU'VE LEARNED: MIXED-ABILITY VERSUS SAME-ABILITY GROUPING IN MATHEMATICS

Liora Linchevski, The Hebrew University of Jerusalem, Israel
Bilha Kutscher, The David Yellin Teachers College, Israel

Abstract. In this article we report on 3 studies in which we investigated the effects of teaching mathematics in a mixed-ability setting on students' achievements and teachers' attitudes. The findings of the first 2 studies indicate that the achievements of students need not be compromised in a heterogeneous setting; on the contrary, the achievements of our average and less able students proved to be significantly higher when compared to their peers in the same-ability classes, whereas highly able students performed about the same. In the 3rd study we show that participating in the project workshops had a positive effect on teachers' attitudes toward teaching in mixed-ability mathematics classes.

Key words: Cooperative learning; Equity/diversity; Grouping for instruction; Longitudinal studies; Quasi-experimental design

THE DEGREE of influence of school grouping methods on the individual student's scholastic achievements is a central issue in educational research. One of the most widespread methods of grouping students in the same grade is ability grouping, either on a subject-by-subject basis (tracking) or for all subjects at once (streaming). Tracking and streaming are widely viewed as the best way to improve the scholastic achievements of all students.

Studies have shown that most teachers have a positive attitude toward ability grouping (Barker-Lunn, 1970; Chen & Addi, 1990; Chen & Goldring, 1994; Guttman, Gur, Kaniel, & Well, 1972; Husén & Boalt, 1967; McDermott, 1976; Oakes, 1985). Many of them justify ability grouping on the basis of the need to adapt class content, pace, and teaching methods to students functioning on different levels (Dar, 1985; Slavin, 1988, 1990; Sørensen & Hallinan, 1986). In the case of mathematics it is also justified by the "nature" of the subject. Mathematics is perceived as "graded," "linear," "structured," "serial," and "cumulative"—making it difficult to work with groups of students with different levels of knowledge and ability. And, indeed, the central issues for supporters of ability grouping relate to "ability to learn mathematics" and "the hierarchical nature of the subject" (Ruthven, 1987). They view students' abilities as the major explanation for differences in their achievements in mathematics (Lorenz, 1982).

Recent research, however, has cast doubt on whether placing students into ability groups is the correct method for dealing with the diversity of abilities. It has generally been shown that the scholastic achievements of students assigned to higher tracks are better than those of students who are judged to have similar abilities but who have been placed in lower tracks. Researchers conducting studies of this sort have concluded that the placement of students in ability groups in and of itself increases the gap between students beyond what would be expected on the basis of the initial differences between them (Alexander, Cook, & McDill, 1978; Gamoran & Berends, 1987; Gamoran & Mare, 1989; Kerckhoff, 1986; Oakes, 1982; Slavin, 1990; Sørensen & Hallinan, 1986).

Reprinted from *Journal for Research in Mathematics Education, 29*, 533–554.

The discouraging results of tracking studies, on the one hand, and evidence of the promising potential of cooperative learning, on the other (Crain & Mahard, 1983; Crain, Mahard, & Narot, 1982; Davidson & Kroll, 1991; Goldring & Eddi, 1989; Wortman & Bryant, 1985; Willie, 1990), have prompted attempts to cope with student diversity within the mathematics classroom. Most of those using these approaches argue that low-ability settings lead to low-quality teaching. Low-quality teaching is characterized by teachers' low expectations; a low-status, nonacademic curriculum; valuable class time spent on managing students' behavior; and most class time devoted to paperwork, drill, and practice. Moreover, the nature and quality of the oral interaction is fundamentally different in low-track and high-track settings (Gamoran, 1993). This last crucial aspect—the role and quality of discussions—is highly emphasized in theoretical approaches that describe learning as an individual process nourished by interpersonal interaction (Bandura, 1982; Carver & Scheier, 1982; Voigt, 1994; Wood & Yackel, 1990). For these theorists the study group is not a mere administrative division but is a crucial component of the learning environment. They suggest that two hypothetically identical students may end up with different mathematical knowledge if they are assigned to two study groups with significantly different participants and styles of interaction. Ability grouping is an obvious case of creating unequal learning groups within the same school. It is thus unsurprising that it has been criticized on this ground and that alternative solutions for dealing with student diversity, such as mixed-ability settings, have been investigated.

Thus, past research has shown that ability grouping results in an increase in the gap between high- and low-ability students beyond that expected on the basis of initial differences between them. What has not been shown is, first, whether this growth in inequality is avoided in mixed-ability settings and, second, whether this gap in achievement (because of tracking) occurs because tracking helps students in the higher ability groups, harms students in the lower ability groups, or because of some combination of the two. In Part I of this article we report on two studies (Study 1 and Study 2) that were designed to address these two questions. In Part II we report on a third study that was designed to examine the effects on teachers' attitudes of teaching in mixed-ability classes. These three studies took place within the framework of a large, ongoing project, Project TAP, in which mathematics is

taught in mixed-ability settings in Israeli junior high schools. In the following section we briefly describe this project.

The TAP Project

The junior high schools participating in the TAP project are comprehensive schools. Each of these schools draws its students from at least two elementary schools that are located in neighborhoods of differing socioeconomic levels. The heterogeneity of each of the classes participating in TAP reflected the heterogeneity of the population of its school.

The major principle of the TAP (*T*ogether and *AP*art) project is to keep a class together as one learning unit while responding to the different needs of the students. This principle does not necessarily mean bringing all the students to the same level of achievement. Instead, it means enabling them to progress to the fullest extent of their abilities through a combination of the following: (a) meaningful instructional activities for cooperative learning by all students throughout the school year in heterogeneous settings whether the whole class or smaller groups—activities henceforth called *shared topics*—and (b) differential instructional activities for cooperative learning by different students according to their abilities and prior achievements in homogeneous settings—henceforth called *differential topics*.

Thus, in each class the teaching was conducted within four major settings: (a) students working in a whole-class setting; (b) students working in small mixed-ability groups; (c) students working in small homogeneous groups; and (d) students working in large homogeneous groups. In the first and last settings teachers played an active role, whereas in the others they were in a supportive role only. Each of these settings was designed to respond to different needs for interaction among the students and between the teacher and the students.

During whole-class discussions the teachers could develop conceptions about what mathematics is; create an appropriate learning atmosphere; and foster essential norms such as listening to classmates, legitimizing errors as part of the learning process, and allowing expression of ideas and tolerance of ambiguity (Davis, 1989; Gooya & Schroeder, 1994). The whole-class discussions established a basis for collaborative dialogues that

Lessons Learned From Research

are known to be a major feature of productive small-group interactions. These discussions also allowed the weaker students to participate, albeit many times passively via "legitimate peripheral participation" (Lave & Wenger, 1991) and "cognitive apprenticeship" (Brown, Collins, & Duguid, 1989), in a challenging intellectual atmosphere.

Justifications for small-group interaction within mathematics classrooms have been presented in many recent papers (Brown et al., 1989; Cobb, 1994; Good, Mulryan, & McCaslin, 1992; Schoenfeld, 1989; Shimizu, 1993; Yackel, Cobb, & Wood, 1991). Cobb (1994) has pointed out that productive small-group interactions involve multivocal interactions, which at first glance seem to require homogeneous grouping. Further, according to Cobb, "Homogeneous grouping … clashes with a variety of other agendas that many teachers rightly consider important, including those that pertain to issues of equity and diversity" (p. 207). Brown et al. (1989) emphasized the cognitive value of collaborative learning via cognitive apprenticeship in heterogeneous groups. We thus chose for our project the strategy of alternately using small homogeneous groups and small heterogeneous groups so that each child was simultaneously a member of two groups (the composition of the groups changed from time to time, depending on the topics, activities, and students' past achievements). The work of the heterogeneous groups focused on the shared topics that met all the requirements of the official curriculum. The homogeneous groups, however, usually dealt with completely different mathematical topics, prepared in accordance with the groups' needs, and sometimes the groups were presented with alternative approaches to the same mathematical topic. In the homogeneous setting opportunities for multivocal interactions were created naturally.

Whenever a teacher felt that a large, specific homogeneous group of students would benefit from the teacher's direct intervention, that setting was used—for example, to better prepare weaker students to be integrated into a planned heterogeneous group activity. Silver, Smith, and Nelson (1995) described activities of this sort as "preteaching." Large homogeneous groups were also used to investigate enrichment topics.

To involve the project teachers in developing the appropriate strategies, tools, and instruments needed to teach effectively in their heterogeneous mathematics classes, we held weekly workshops in which all teachers participated. For example, activities were prepared for different ability levels or to encourage interactions in heterogeneous groups. An equally important aspect of our workshop meetings was discussion and sharing of problems that had arisen in the teachers' classes that week.

The present studies do not specifically examine any particular aspects of the TAP project. The purpose of the project description is to give the reader some flavor and familiarity with the project because the schools involved in these studies were all project schools.

In the next section we report on two studies in which we examined how learning mathematics in mixed-ability settings affected students' achievements.

PART I: STUDENTS' ACHIEVEMENTS

Rationale for Our Research Questions and Study Design

In Study 1 and Study 2 we compared achievements of students studying in mixed-ability versus same-ability systems to determine (a) which format leads to greater achievements on the part of the students and (b) specifically, which system leads to greater achievements for the better students, the weaker students, and the intermediate ones, when parallel levels for each of these systems are compared.

The appropriate design for examining these questions is an experiment involving the random assignment of classes of students to either heterogeneous or homogeneous classes. In the heterogeneous classes the students are hypothetically assigned to ability-group levels, whereas in the homogeneous classes the students actually study according to the assigned ability levels. However, the difficulty in performing random experiments in the educational system and the methodological problems associated with post hoc comparisons between schools with and without ability grouping (for a review see Slavin, 1990) led to a less ambitious design, one that compared ability-group levels within schools. In the latter type of study, one investigates whether the gap between better and weaker students after placement in ability groups for a certain period of time differs from the gap expected on the basis of initial differences (Kerckhoff, 1986; Oakes, 1982). In this type of study the main methodological problem is to separate the two effects that might influence final achievements: the effect of belonging to groups at different levels and the effect of the initial differences between the students placed in these groups (Cahan, Linchevski, & Ygra, 1992).

The methodological problem can be overcome if students are divided into group levels by establishing agreed-upon cutoff points based on a measure of ability, previous achievements in the subject matter at hand (henceforth called the pretest), or both. Using information on the cutoff points and each student's group level and pretest score, one can see the variance among the student scores on a common achievement test some time later (henceforth called the posttest) as the sum of two effects: (a) the effect of the initial differences among the students and (b) the effect of the group levels in which students worked. These two effects can be disentangled by means of a regression discontinuity design (Cook & Campbell, 1979). The effect of initial differences is estimated by the regression line of the posttest on the pretest within each group level, whereas the effect of the group level is estimated by the discontinuity between the regression lines of consecutive group levels. This research design is known as the quasi-experimental regression discontinuity design. Figure 1 shows an example of all possible combinations of the two effects.

Students close to a given cutoff point on either side can be seen as identical from the viewpoint of the selection criterion. Figure 7.1a is a hypothetical example of "no grouping effect": After a period of treatment no gap has been created at the cutoff points; that is, after a period of treatment students on either side of and close to a cutoff point had similar scores. Thus the variance among the students' scores is due only to the initial differences among the students. Figures 7.1b and 7.1c are hypothetical examples in which gaps have been created at the

cutoff points. Thus the variance among the students' scores is due to the initial differences among the students and to a grouping effect. In Figure 7.1b students in the higher group level gained more than similar students in the lower group level, whereas in Figure 7.1c the opposite occurred.

A design of this sort was used successfully by Abadzi (1984, 1985) in the United States for investigating the effect of streaming in elementary schools and in Israel by Cahan and Linchevski (1996) for investigating the effect of tracking in mathematics in junior high schools. In the latter investigation the findings clearly indicated that the differences among the scholastic achievements of the students at the different group levels at the end of the first and the third years of junior high school were greater than would be predicted by the data at the time of placement. Moreover, in most of the schools, after 3 years, the effect of the group level was greater than the effect of the initial differences among the students.

Study 1

Because widening the gap between stronger and weaker students might occur in heterogeneous settings as well as homogeneous settings, the results reported above have no clear bearing on the comparative benefits of homogeneous and heterogeneous educational settings (Cahan et al., 1992; Linchevski, Cahan, & Dantziger, 1994). Study 1 was designed to answer the following question: Is the gap between better and weaker students learning together in mixed-ability settings for a certain period of time different from the gap that would be expected on the basis of initial differences between the two groups? We used the regression discontinuity design to investigate this question. In the study we also compared the results with those reported in the ability-grouping study of Cahan and Linchevski (1996). This comparison was possible because identical research designs were used in the two studies.

Our conjecture was that in the schools that participated in the TAP project, no gaps between students would be created beyond the one expected on the basis of initial differences among them. This outcome was expected because the project was designed using the theoretical considerations and previous research results reported in this article (e.g., Brown et al., 1989; Cobb, 1994; Gamoran, 1993; Lave & Wenger, 1991; Schoenfeld, 1989).

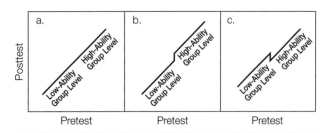

Figure 7.1. An example of the various combinations of the values of the two effects: a. Lack of grouping effect. b. Favoring high-ability group. c. Favoring low-ability group.

Lessons Learned From Research

Design

In Study 1, the unit of analysis was a school. This choice was made for several reasons. First, the fact that the effects were estimated separately for each of the schools investigated actually constitutes independent replications of the study. Second, we could compare this study with the earlier ability-grouping study by Cahan and Linchevski (1996) because the same unit of analysis was used in the two studies. Last but not least, the choice of a school as a unit of analysis is an improvement on earlier studies that compared the gap between heterogeneous and homogeneous settings. Other earlier studies calculated effects in a pooled sample of schools, hence obtaining results that might have hidden the variability in mathematics achievement by aggregation of schools to the pool level.

Because teaching actually took place in heterogeneous settings in this study, the application of the regression discontinuity design required a hypothetical division of the students into the various group levels as if these students were actually going to study in separate ability groups. At the beginning of the seventh grade, each school therefore assigned each student his or her ability-group level according to the school's previous tracking policy, although in effect these students studied mathematics in heterogeneous settings. This assigning procedure was done without the knowledge of the students' mathematics teachers. The hypothetical group level and the placement scores (the pretest) served as the independent variables, whereas the achievement test scores in mathematics after 1 year and after 2 years—at the end of the seventh grade and at the end of the eighth grade (the posttests)—served as the dependent variables. On the basis of these variables, the effects of grouping and initial differences on achievement were calculated separately for each school. For a detailed description of these calculations see the appendix.

Sample

All 1730 seventh-grade students in the 12 Israeli junior high schools that participated in the TAP project were tested (posttest) at the end of the seventh grade. We have complete data for 1629 of these students.

In 4 of the 12 schools, the students (389 students) were tested at the end of the eighth grade as well.

Tests

Achievements in mathematics were measured by tests constructed according to the topics covered in the schools, as detailed in the national mathematics curriculum (first- and second-year algebra, problem solving, and geometry). The contents of the seventh- and eighth-grade tests were confirmed with the schools involved, and the tests were validated by experts and by the General Inspector for Mathematics Teaching in the Israeli Ministry of Education. The size of the research population and the nature of the study determined to a large extent the type of test that could be administered to the students. We are aware that these tests were traditional in form. However, the questions were not multiple-choice but instead were open-ended, allowing for more flexible types of questions to be asked. To control for between-school differences in mean achievement levels and test-score variance, the scores on pretests and posttests were standardized separately for each school, with a mean of 0 and standard deviation of 1.

Results

Results at the end of the seventh grade. The measures of effects of the hypothetical ability groupings on the students' achievements in mathematics at the end of the seventh grade are presented separately for each school (see Table 7.1).

Table 7.1
The Results of the Regression at the End of the 7th Grade

School	No. of groups	Hypothetical-differences effect (α_H)[a]	Hypothetical-differences effect *p* value for *F*-test	Initial-differences effect (α_P)[a]
1	4	-0.36	.37	2.75
2	3	-0.42	.18	3.17
3	3	-0.26	.54	2.30
4	3	-0.60 *	.04	2.75
5	4	-0.12	.64	2.71
6	4	-0.87 *	.03	3.45
7	3	0.10	.75	2.78
8	4	0.09	.75	3.15
9	3	-0.34	.42	3.11
10	3	-0.40	.20	3.57
11	3	0.26	.59	2.20
12	3	0.10	.71	2.90

[a] For details, see the the appendix.
* $p < .05$.

A "negative" grouping-effect measure (e.g., –0.36, see Table 7.1) means that a gap that was not expected on the basis of initial differences was created at a cutoff point. This negative grouping effect means that students close to a cutoff point gained more on the average if they were in a higher hypothetical group than if they were in the next lower group (see Figure 7.1b). A "positive" grouping effect means that students close to a cutoff point gained more being hypothetically part of a lower ability group than of the next higher group (see Figure 7.1c). Zero effect measure means grouping had no effect—neither increasing nor decreasing the gap (see Figure 7.1a). As can be seen in Table 7.1, the effects were not uniform and in 10 of the 12 schools were nonsignificant at the p = .05 level. In 8 of the 12 schools the effect was negative, whereas in the others it was positive. The size of the effect measure was different in different schools, ranging from +0.26 SD to –0.87 SD, with a median of –0.31 SD.

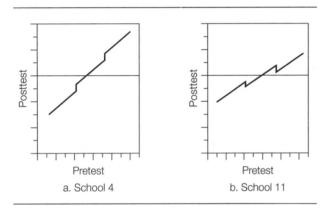

Figure 7.2. Representative cases of discontinuity in posttest and pretest grades.

Figures 7.2a and 7.2b show two examples of the regression discontinuity of the posttest on the pretest. School 4 is a case in which the effect was in the direction of increasing the variance. The difference between the better and the weaker students was 3.35 standard deviation units (2.75 – [–0.60] = 3.35; see appendix) at the end of the year.

In School 11, the difference between the two groups at the end of the year was less than would be expected on the basis of the initial differences—a gap of 1.94 standard deviation units (2.20 – [+0.26] = 1.94; see appendix).

A comparison of the hypothetical-grouping effect with the initial-differences effect (see Table

7.1) in each of the schools shows that the former was very small in all schools, relative to the initial-differences effect. Because the hypothetical-grouping effect was nonsignificant in 10 out of the 12 schools and the trend was not uniform—that is, the effect was positive in some of the schools and negative in others—the conclusion is that in 10 of the schools there was no effect and in 2 schools there was a negative effect.

Results at the end of the eighth grade. Four of the 12 schools that participated in the study maintained heterogeneous classes in the eighth grade as well. Table 7.2 shows the hypothetical-grouping effects and the initial-differences effects at the end of 2 years of such studying in these four schools.

Table 7.2

The Results of the Regression at the End of the 8th Grade

School	No. of groups	Hypothetical-grouping effect $(\alpha_H)^a$	*p* values for *F*-test	Ability effect $(\alpha_P)^a$
4	3	–0.52*	.04	2.70
6	4	–0.36	.40	3.00
7	3	–0.24	.49	2.80
11	3	0.10	.70	2.70

[a] For details, see the the appendix.
* *p* < .05.

In three of the four schools there was a negative treatment effect—that is, the variance at the end of the 2 years was greater than would be expected on the basis of the placement data, whereas the opposite effect was found in the remaining school. In Schools 4 and 6 the effect at the end of the seventh grade had been negative, and it continued to be negative at the end of the eighth grade, but its absolute value at the end of the second year was smaller than it had been at the end of the first. In School 7 the effect changed from positive at the end of the seventh grade to negative at the end of the eighth grade, whereas in School 11 it remained positive, although its absolute value decreased. In three of the four schools the effect was not significant. Only in School 4 was the effect significant; the effect had been significant there at the end of the seventh grade as well. Generally speaking, inasmuch as the effects were not significant (other than in School 4), the conclusion is that in heterogeneous classroom instruction the differences in achievement are explained mainly by the initial differences.

Comparison between the present study and the ability-group study

The main difference between the results of this study and those of the same-ability-group study previously described (Cahan & Linchevski, 1996) is the near absence of any treatment effect in this mixed-ability-group study and the presence of a unidirectional treatment effect in all the schools in the previous same-ability-group study.

In the same-ability-group study the variance among the students at the end of the seventh grade was greater than would be expected according to the placement data in each of the schools. Moreover, the grouping effect increased over the years in all the schools. The grade of a student in a higher ability group was always higher than the grade he or she would have received if he or she had hypothetically been placed, with the same initial data, in the next lower ability group. Thus it seems that in a tracking system the achievements of students close to the cutoff points are largely dependent on their being arbitrarily assigned to a lower or higher group level. In the present study, in which the students attended heterogeneous classes, there was no significant effect in 10 of the schools at the end of the seventh grade. In the other two schools (Schools 4 and 6) there was a significant effect, and it was in the same direction as in the same-ability study. At the eighth-grade level, three of the cases showed no significant effect, whereas in the fourth, there was a significant difference but with a smaller absolute value than that found at the end of the seventh grade. One important difference between the two studies, however, is that in the same-ability-grouping study the students were tested at the end of 1 year and 3 years of grouping, whereas in the mixed-ability study they were tested at the end of 1 year and 2 years.

Study 2: Another Perspective

In Study 2 we compared the mathematical achievements (actual grades) of students placed in same-ability classes with those of students placed in mixed-ability classes to investigate our second research question: Which of the two systems—placement in heterogeneous classes as described earlier in the project description or placement in same-ability classes—leads to greater student achievement? Moreover, which system leads to greater achievements for the better students, the intermediate students, and the weaker students, when parallel levels in each of these systems are compared?

Our conjecture was that the achievements of lower and intermediate level students who learned in heterogeneous settings would be higher than those of students in parallel levels who learned in homogeneous settings. This outcome was expected because the students in the TAP classes had the advantage of participating in a rich learning environment that included legitimate peripheral participation (Lave & Wenger, 1991) and cognitive apprenticeship (Brown et al., 1989) via cooperative learning.

With respect to the highest level students, no differences were expected for two reasons: (a) In Gamoran's (1993) analysis, the gap found in the ability-grouping system emanates more from the weaker students' loss than from the stronger students' gain because of the qualitative differences in the students' learning environments; (b) the learning environments in our heterogeneous classes, experienced by all students and in particular the strongest ones, incorporate most of the positive factors encountered by the highest level ability groups in the tracking systems.

Design

In a junior high school not associated with TAP the mathematics faculty considered the possibility of joining the project because they had been quite unhappy with their instruction in the lower tracks. However, they considered same-ability grouping the only fair, effective way to deal with student heterogeneity. There was a conflict between this belief, on the one hand, and the project's stated benefits of learning in heterogeneous classes, on the other hand. After a long process of deliberation that included meetings, discussions, and reading some of the relevant research literature, the mathematics teachers decided to participate in a "real experiment" for which they obtained the parents' agreement. Thus, this school was selected for this study.

For the study we used a random experimental design, with the class as the unit of analysis. At the beginning of the seventh grade all the students were new to the school and were randomly assigned to four mixed-ability homeroom classes. (All content areas other than mathematics were taught, as in the past, in these homeroom settings.) Thereafter, using the same procedure the school had always used for tracking in mathematics, all students were assigned to one of three ability-group levels for mathematics. Two of the large mixed-ability homeroom classes were tracked into three smaller separate, same-ability mathematics classes according to the assignment procedure, and the other two homeroom classes studied mathematics in their original mixed-ability homeroom classes. Thus, there were five mathematics classes in all: two mixed-ability classes and three same-ability classes. In the two mixed-ability classes, the same tracking procedures were used to assign students to hypothetical ability groups. This hypothetical assignment was disclosed neither to the teachers nor to the students. Teachers were randomly assigned to the different mathematics classes, and teachers of both kinds of groups were involved in weekly workshops: the mixed-ability-group teachers' workshop was led by a TAP counselor, and the same-ability-group teachers' workshop was led by the school mathematics coordinator. Whereas the project workshop concentrated on discussions and activities appropriate for the heterogeneous classes, the same-ability workshop concentrated on discussions and activities appropriate for the same-ability levels.

Students remained in the groups for 2 years. At the end of the eighth grade, achievement tests were administered to all students. Two alternatives had been discussed: One was giving all students the same test regardless of their placement; the other alternative was writing three different tests for the three different ability-group levels, with students in the true same-ability classes and students of the equivalent hypothetical level in the mixed-ability classes being given the same test. This alternative was discussed to reduce anxiety of students who had learned in the lower same-ability classes and had been accustomed to tests specially prepared for their levels. In the end

it was decided that both forms of testing would be used. The questions for these tests were proposed by the teachers and the mathematics coordinator, and the final versions were written by a representative of the Ministry of Education.

Results

Table 7.3 displays the average scores for each group level on the final achievement tests for the differential tests and for the common test. The data have been presented for same-ability groups and for mixed-ability groups. *T*-tests were used to compare, for each level, the achievements of the students in the mixed-ability and same-ability classes. The average scores of high-level students in the same-ability classes were higher (but not significantly) than those of the students in the mixed-ability classes on both versions. The average scores of intermediate- and low-level students in the same-ability grouping system were significantly lower than those in the mixed-ability system on both versions. Moreover, it seemed that most low-level students in the same-ability classes were unable to answer the questions on the common test inasmuch as they handed in almost empty test papers, whereas the equivalent students in the heterogeneous classes scored an average of 54%. The students in the mixed-ability classes who had been hypothetically assigned to the intermediate and low tracks found the differential tests, written for the actual lower tracks, relatively easy; they were accustomed to much higher demands and expectations. Thus our hypotheses concerning all levels were confirmed.

An analysis of the test papers showed that the better students in the mixed-ability classes lost points for formal presentation and notation, such as the symbolic notation of truth set or the formal

Table 7.3

Achievements (means in Percentages) in Mathematics at the End of 8th Grade

Tests	Same-ability groups			Mixed-ability groups		
	High	Intermediate	Low	High	Intermediate	Low
Differential test						
Mean	85	64	55	82	80*	78*
SD	7.8	5.6	6.2	7.5	4.3	5.1
n	33	27	14	35	26	15
Common test						
Mean	88	41	—[a]	85	65*	54
SD	8.1	5.1	—	6.9	6.1	3.9
n	33	27	14	35	26	15

[a] Because many of these students did not complete the test, we could not do a *t*-test, but because the mean would have been exceedingly low, we assume that it would be significantly different from 54, the score of the low-ability students in the mixed-ability group.

*$p < .05$ (significant *t*-test value between the same-ability mean and the mixed-ability mean).

presentation of geometric proofs, whereas high-level students in the same-ability classes were quite competent in this area. The teachers of the mixed-ability classes confirmed that some symbolic notations and formal presentations, which traditionally constitute a major problem for average students, were given less attention in their classes.

PART II: TEACHERS' ATTITUDES

Another purpose of our research project was to examine the attitudes of the TAP teachers toward teaching in heterogeneous classes and the effects of workshops on these attitudes. Study 3 was designed for this purpose.

Past research (Chen & Addi, 1990; Chen & Goldring, 1994; Oakes, 1985) has shown that teachers have conflicting attitudes about ability grouping. Ideologically they favor diversity; practically they support ability grouping. Dar (1985) showed that teachers who work in heterogeneous settings within a school that actively supports and maintains such classes have a more positive attitude about the effectiveness of heterogeneity than their colleagues who teach in schools with a policy of tracking students into homogeneous classes. Thus, direct experience of teaching in a heterogeneous setting within a supportive framework might have a positive effect on teachers' perspectives about non-tracking of students. Dar (1985) emphasized, however, that even these teachers still maintained a negative attitude toward teaching mathematics and English in heterogeneous classes. Schools with teachers who have succeeded in teaching effectively in heterogeneous classrooms were reported to have involved the teachers in the ideological development and implementation of a commitment to education in diverse classrooms (Wheelock, 1992). Professional training of the teachers is another factor crucial to successful implementation if tracking is to be eliminated (Gamoran, 1992). Thus, in planning the project-workshop guidelines we took into consideration the aspects mentioned above. All the project teachers participated in weekly workshops in which discussions evolved concerning the project's rationale and ideological basis. All project workshops were led by TAP project counselors who had participated in a special course developed for them. Thus all project teachers received approximately the same inservice training. The project teachers developed an awareness of the different needs of the students and were involved in constructing instruments, tools, and appropriate strategies for cooperative and differential teaching in their heterogeneous mathematics classes. Preparation for differential teaching included gaining familiarity and proficiency with differential classroom strategies and class organization, preparing relevant activities, preparing alternative assessment tools, and the like. An equally important purpose of our workshop meetings was the opportunity given for discussing and sharing problems that had arisen in the teachers' classes that week. Frustrations, successes, and failures were all exposed and possible solutions were sought by the group; thus the teachers received essential support and practical solutions for their needs.

Study 3

In Study 3 we examined the attitudes of all project teachers who taught mathematics in heterogeneous settings while participating in the project, investigating the following research question: What are the attitudes of the teachers participating in the TAP project toward teaching mathematics in heterogeneous classes, and how do the project workshops affect these attitudes?

Design

The target population for this study was the group of all the project teachers, 58 teachers from demographically diverse regions. All answered a written questionnaire. Shortly thereafter individual oral interviews were conducted with 5 of these teachers. We defined (teacher) seniority as the number of school years the teacher had participated in the TAP project workshops. For example, if a teacher participated in both a seventh- and eighth-grade workshop during one school year, he or she accumulated two workshop-years. The teachers' seniority in the project varied from one to five workshop-years. Our conjecture was that positive attitudes toward teaching mathematics in heterogeneous settings would be directly related to the teachers' number of workshop-years. This outcome was expected because the project workshop guidelines had incorporated the suggestions emanating from the relevant research reported in this section.

The survey questionnaire contained 51 items. All items were constructed as statements, to be evaluated according to a 5-point Likert-type scale. Scores lower than 3 represented support for heterogeneity in terms of attitudes, workshop contribution, and absence of instructional difficulties.

The questionnaire comprised four parts: The first part was subdivided into four topic factors and the second part into two topic factors; the third and fourth parts had only one topic factor each, for a total of eight topic factors.

1. Teachers' attitudes toward children's learning in heterogeneous classes included 14 items. This part of the questionnaire included four factors: (a) Factor 1: Affective Impact of Heterogeneity (e.g., "Heterogeneous classes rid weak pupils of feelings of inferiority"); (b) Factor 2: Cognitive Effects of Heterogeneity (e.g., "Studying in a heterogeneous class challenges the low-ability students"); (c) Factor 3: Equality of Educational Opportunity Selection (e.g., "Learning in ability groups increases the gap between the high- and low-ability students"); and (d) Factor 4: Reliability and Validity of Educational Selection (e.g., "It is possible to place students accurately into ability groups").

2. Instructional difficulties in heterogeneous classes included 12 items divided topically into two factors: (a) Factor 5: Solutions by Training meant instructional difficulties that could be resolved by appropriate training (e.g., "I lack knowledge of mathematical instructional methods needed to teach mathematics in a heterogeneous class") and (b) Factor 6: External Constraints meant instructional difficulties such as class size, inadequate teaching or learning materials, and so on (e.g., "It is impossible to teach mathematics in a heterogeneous class with 40 pupils").

3. The importance of various items in differential teaching in heterogeneous classes included 8 items related to Factor 7: Item Importance in Differential Teaching (e.g., "Pupils assisting each other in learning is important to me in differential teaching").

4. The workshop's contribution to the teachers included 17 items related to Factor 8: Workshop Contribution (e.g., "The workshop allows me to raise and solve instructional problems that come up in class").

Relevant items relating to teachers' attitudes toward children's learning in heterogeneous classes were adapted from an attitude inventory (Dar, 1985). The items relating to instructional difficulties in heterogeneous classes were in part extracted from a survey conducted by Chen, Kfir, and Addi (1990) and in part derived from interviews with teachers in heterogeneous mathematics classes.

Items for Factors 7 and 8 were based on interviews with project counselors and former project teachers. The questionnaire was pilot tested among those teachers who had participated in the project but had since left the schools for various reasons. Internal reliability of the questionnaire was measured using Cronbach's coefficient alpha. Reliability coefficients ranged between 0.73 and 0.88. This result confirmed the validity of the distinction for the eight factors.

For statistical reasons emanating from the sample size, the teachers were grouped into three seniority groups: (a) Seniority 1 consisted of teachers with experience of one workshop-year, (b) Seniority 2 & 3 consisted of teachers with two or three workshop-years, and (c) Seniority 4+ consisted of teachers with four to seven workshop-years' experience.

Results

The measures of attitudes toward teaching in heterogeneous classes are shown in Figure 7.3. The scores showed, generally, that the attitudes of project teachers with more seniority were more positive toward student learning in a heterogeneous class than the attitudes of new project teachers.

Using analysis of variance, we compared the three seniority groups. The biggest difference found was with respect to cognitive effects on the students when learning in a heterogeneous class (p = .004, cognitive effects of heterogeneity, Factor 2, Figure 7.3). Whereas the novice project teachers had reservations (x = 3.16), the teachers with four or more years' seniority were more confident of positive effects of heterogenization (x = 2.16). These findings contradict Dar's (1985) results, which showed that even those teachers who were currently teaching in heterogeneous classes and had reservations favored ability grouping in mathematics (and in English). It might thus be assumed that participation in the workshop training program positively affected the teachers' attitudes.

Using the fact that scores lower than 3 represent support for heterogeneity, we can say that all three groups of project teachers agreed that heterogeneous grouping enhanced the weaker students' self-images and motivation (Factor 1) and all groups disagreed that the assignment of all students to a certain ability group was either reliable or valid (Factor 4). Significant differences (p = .03) appeared among the groups of teachers in their

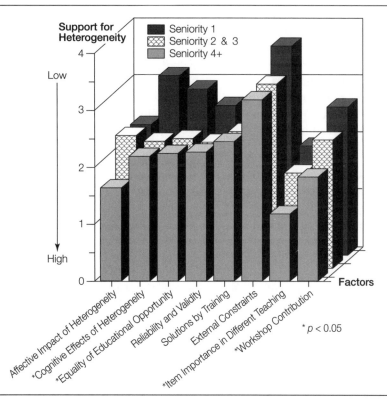

Figure 7.3. Support for heterogeneity according to teacher seniority groups (by questionnaire factors).

appraisals of the effect of equal educational opportunity on all students. Whereas the novice project teachers were but slightly positively inclined, the teachers with seniorities of 2 and 3 or 4+ were far more positive of the effects (Factor 3). All groups of project teachers agreed that they were capable of teaching mathematics in heterogeneous classes (Factor 5). However, large classes, the lack of necessary equipment, and the curricular demands dictated by higher grade levels were difficulties with which all project teachers were unhappy (Factor 6). The findings presented in Figure 7.3 show that not all the problems were solved through the workshops. Teacher training-related problems were resolved; difficulties stemming from external constraints were not.

All items included as items important in differential teaching were found to be very important to all project teachers with significant differences among them ($p = .002$, Factor 7); the teachers with greater project seniority found these items of greater consequence to their lessons. Seemingly, experience in differential teaching and, perhaps, positive pupil feedback raise the teachers' consciousness of the usefulness of these items. A simi-

lar trend was found for items associated with the workshop contribution ($p = .034$, Factor 8); the project teachers of greater seniority felt that the workshop had better equipped them for the task of teaching in a heterogeneous class more so than did the novices. Thus, the more experienced project teachers seem to echo Gamoran's (1992) suggestion that professional training is essential for the successful implementation of teaching in heterogeneous (mathematics) classes.

The question arises whether these teachers feel capable of coping with the task of teaching mathematics in diverse classrooms without the continuous support of regular workshop meetings. To find out we examined the teachers' responses in the oral interviews. All the teachers felt the need for regular meetings in which lessons would be discussed and planned and common problems resolved, in effect emulating the format of the project workshop.

The picture that emerges is that continuous intercollegial support seems to be crucial for the success of implementing a program that requires fundamental changes in instructional methods.

Conclusions and Discussion

In this article we report on three studies of teaching mathematics in mixed-ability and same-ability settings and the effects of the settings on students' achievements and on teachers' attitudes. These studies took place within the framework of the TAP project. The reported results, to a great extent, support our conjectures.

We first examined the ways in which teaching in mixed-ability mathematics settings affects students' achievements. In Study 1 we investigated whether studying in mixed-ability classes would prevent formation of a gap (usually found when students are grouped by ability) between high- and low-ability students greater than that expected on the basis of the initial differences between them. In Study 2 we compared the effects of mixed-ability and same-ability grouping on the mathematics performance of students classified as having high ability, intermediate ability, and low ability. In Study 3 we examined how teaching in mixed-ability classes affects teachers' attitudes.

The results of Study 1 showed that after 1 year, in 10 of the 12 schools investigated, there was no significant change in the achievement differences among students of different ability levels as a result of using mixed-ability grouping. The 2 remaining schools showed a statistically significant increase in this gap after 1 year. By the end of 2 years the effects in both of these schools had been decreased; in only 1 school was the effect still significant. Thus we may conclude that within the TAP schools the added gap that is created in a tracking system was nearly nonexistent.

The results of Study 2 showed that placement of students in mixed-ability mathematics classes was not detrimental to their achievements when compared to achievements of students of similar ability levels who had learned in separate same-ability classes. On the contrary, the average and weaker students' achievements showed significant gains, whereas the loss in achievements of the stronger students was negligible. Demonstrating this result was one of the project developers' main goals.

By integrating the results of Studies 1 and 2, we might conclude that an increase in the gap, due to learning in the tracking system, emanates mainly from the loss in the weaker students' achievements instead of from the stronger students' gains. In Study

1 we have shown, generally, that in mixed-ability classes the gap did not increase nor were achievements significantly impaired. In Study 2 the comparison between the achievements of the mixed-ability students and their same-ability counterparts indicates that the achievements of the average and lower ability groups in the mixed-ability classes were higher. We may, then, conclude that in our case all levels progressed reasonably well. As such, we may infer that the increase in the gap due to learning in the same-ability classes emanates mainly from the loss for the students in the lower ability levels instead of from gain for the stronger ones. Better understanding of cognitive differences among students requires further investigation. By using alternative assessment tools that necessarily require a smaller research population, researchers can take a closer look at the students' thinking processes. Such alternative analyses may explain differences among students of different levels in the two systems.

In Study 3 we investigated the TAP-project teachers' attitudes. It must be remembered that the schools involved in the TAP project were not randomly sampled. Most principals were interested in the program; perhaps they felt increasingly skeptical toward ability grouping because of difficulties encountered in the lower ability levels or because they really believed that this program offered a chance to achieve equity. It does not necessarily follow that all participating teachers wanted to participate in this project. On the contrary, most of the teachers did not initially believe it was possible to successfully implement a mixed-ability mathematics program.

Study 3 results suggest that TAP-project participation had a positive effect on teachers' attitudes toward teaching in mixed-ability mathematics classes. Those teachers of higher project seniority consistently felt more positive than project newcomers about teaching in mixed-ability mathematics classes. They believed that they were capable of conducting mathematics classes in a manner that would not be detrimental to students at any ability level. They also felt confident that they had acquired tools to challenge all levels of students in a heterogeneous class. Our findings, in contrast with Dar's (1985) results, indicate that mathematics teachers can develop positive attitudes toward teaching in mixed-ability classes.

Thus we learn from the results of Study 3 that it is possible to teach mathematics in a heterogeneous setting to the satisfaction of the teachers involved. All the teachers felt that their success was to some extent dependent on continual support of a workshop type of framework, supporting Gamoran's (1992) suggestions for successfully implementing innovative programs.

Our studies indicate that it is possible for students of all ability levels to learn mathematics effectively in a heterogeneous class, to the satisfaction of the teacher.

REFERENCES

Abadzi, H. (1984). Ability grouping effects on academic achievement and self-esteem in a southwestern school district. *Journal of Educational Research*, 77, 287–292.

Abadzi, H. (1985). Ability grouping effects on academic achievement and self-esteem: Who performs in the long run as expected. *Journal of Educational Research*, 79, 36–39.

Alexander, K. L., Cook, M., & McDill, E. L. (1978). Curriculum tracking and educational stratification. *American Sociological Review*, 43, 47–66.

Bandura, A. (1982). Self efficacy mechanism in human agency. *American Psychologist*, 37, 122–147.

Barker-Lunn, J. C. (1970). *Streaming in the primary school*. Slough, England: NFER.

Brown, J. S., Collins, A., & Duguid, P. (1989). Situated cognition and the culture of learning. *Educational Researcher*, 18 (1), 32–42.

Cahan, S., & Linchevski, L. (1996). The cumulative effect of ability grouping on mathematical achievement: a longitudinal perspective. *Studies in Educational Evaluation*, 22 (1), 29–40.

Cahan, S., Linchevski, L., & Ygra, N. (1992). *Ability grouping and mathematical achievements in Israeli junior high schools*. Jerusalem: Hebrew University, School of Education, The Institute for Research NCJW: Research for Innovation in Education.

Carver, C. S., & Scheier, M. F. (1982). Control theory: A useful conceptual framework for personality— social, clinical, and health psychology. *Psychological Bulletin*, 92, 111–135.

Chen, M., & Addi, A. (1990). *School leaders and teachers' attitudes towards teaching in secondary schools* (Research report). Tel Aviv, Israel: Tel Aviv University, School of Education.

Chen, M., & Goldring, E. B. (1994). Classroom diversity and teachers' perspectives of their workplace. *The Urban Review*, 26 (2), 57 –73.

Chen, M., Kfir, D., & Addi, A. (1990). *Hayeda al migvan shitot hora'a ve'hashimush bahen bevatei sefer yesodi'im ubehativot habena'im: Mimtza'ei seker* [The knowledge of diverse teaching methods and the use of them in primary and junior high schools: Survey findings]. Tel Aviv, Israel: Tel Aviv University, School of Education, The Unit for Sociology of Community Education. Petah Tikva Project (in Hebrew).

Cobb, P. (1994). A summary of four case studies of mathematical learning and small-group interaction. In J. P. da Ponte & J. F. Matos (Eds.), *Proceedings of the eighteenth international conference for the Psychology of Mathematics Education* (Vol. 2, pp. 201–208). Lisbon, Portugal: University of Lisbon.

Cook, T. D., & Campbell, D. T. (1979). *Quasi-experimentation: Design & analysis issues for field settings*. Chicago: Rand McNally.

Crain, R. L., & Mahard, R. E. (1983). The effect of research methodology on desegregation-achievement studies: A meta-analysis. *American Journal of Sociology*, 88, 839–854.

Crain, R. L., Mahard, R. E., & Narot, R. E. (1982). *Making desegregation work: How schools create social climates*. Cambridge, MA: Ballinger.

Dar, Y. (1985). Teachers' attitudes toward ability grouping: Educational considerations and social and organizational influences. *Interchange*, 16 (2), 17–38.

Davidson, N., & Kroll, D. L. (1991). An overview of research on cooperative learning related to mathematics. *Journal for Research in Mathematics Education*, 22, 362–365.

Davis, R. B. (1989). The culture of mathematics and *the culture of schools. Journal of Mathematical Behavior, 8*, 143–160.

Gamoran, A. (1992). Is ability grouping equitable? *Educational Leadership*, *50* (2), 11–17.

Gamoran, A. (1993). Alternative uses of ability grouping in secondary schools: Can we bring high-quality instruction to low-ability classes? *American Journal of Education*, *102*, 1–22.

Gamoran, A., & Berends, M. (1987). The effects of stratification in secondary schools: Synthesis of survey and ethnographic research. *Review of Educational Research*, *57*, 415–435.

Gamoran, A., & Mare, R. D. (1989). Secondary school tracking and educational inequality: Comparison, reinforcement, or neutrality? *American Journal of Sociology*, *94*, 1146–1183.

Goldring, E. B., & Eddi, A. (1989). Using meta-analysis to study policy issues: The ethnic composition of the classroom and academic achievement in Israel. *Studies in Educational Evaluation*, *15*, 231–246.

Good, T. L., Mulryan, C., & McCaslin, M. (1992). Grouping for instruction in mathematics: A call for programmatic research on small-group processes. In D. A. Grouws (Ed.), *Handbook of research on mathematics teaching and learning* (pp. 165–196). New York: Macmillan.

Gooya, Z., & Schroeder, T. (1994). Social norm: The key to effectiveness in cooperative small groups and whole class discussions in mathematics classrooms. In J. P. da Ponte & J. F. Matos (Eds.), *Proceedings of the eighteenth international conference for the Psychology of Mathematics Education* (Vol. 3, pp. 17–24). Lisbon, Portugal: University of Lisbon.

Guttman, Y., Gur, A., Kaniel, S., & Well, D. (1972). *Hashpa'at hakbatzat talmidim (lefi ramot cosher) al hesegim limudi'im vehitpat'hoot psicho-hevratit* [The effects of ability grouping on learning achievements and psycho-social developments]. Jerusalem: Szold Institute.

Husén, T., & Boalt, G. (1967). *Educational research and educational change: The case of Sweden*. Stockholm: Almqvist & Wiksell.

Kerckhoff, A. C. (1986). Effects of ability grouping in British secondary schools. *American Sociological Review*, *51*, 842–858.

Lave, J., & Wenger, E. (1991). *Situated learning: Legitimate peripheral participation*. Cambridge, UK: Cambridge University Press.

Linchevski, L., Cahan, S., & Dantziger, I. (1994). *The accumulating effect of grouping on the achievements in mathematics* (Research Report). Jerusalem: Hebrew University, School of Education, The Institute for Research NCJW: Research for Innovation in Education.

Lorenz, J. H. (1982). On some psychological aspects of mathematics achievement assessment and classroom interaction. *Educational Studies in Mathematics*, *13*, 1–19.

McDermott, J. W. (1976). The controversy over ability grouping in American education, 1916 – 1970 (Doctoral dissertation, Temple University, 1976). *Dissertation Abstracts International*, *37*, 2026–2027A. (University Microfilms No. 76–22056)

Oakes, J. (1982). The reproduction of inequality: The content of secondary school tracking. *The Urban Review*, *14*, 107–120.

Oakes, J. (1985). *Keeping track: How schools structure inequality*. New Haven, CT: Yale University Press.

Ruthven, K. (1987). Ability stereotyping in mathematics. *Educational Studies in Mathematics*, *18*, 243–253.

Schoenfeld, A. H. (1989). Ideas in the air: Speculations on small group learning, environmental and cultural influences on cognition and epistemology. *International Journal of Educational Research*, *13* (1), 71–88.

Shimizu, Y. (1993). The development of collaborative dialogue in paired mathematical investigation. In I. Hirabayashi, N. Nohda, K. Shigematsu, & F. -L. Lin (Eds.), *Proceedings of the seventeenth international conference for the Psychology of Mathematics Education* (Vol. 3, pp. 73–80). Tsukuba, Japan: University of Tsukuba.

Silver, E. A., Smith, M. S., & Nelson, B. S. (1995). The QUASAR Project: Equity concerns meet mathematics education reform in the middle school. In W. G. Secada, E. Fennema, & L. B. Adajian (Eds.), *New directions for equity in mathematics education* (pp. 9–56). New York: Cambridge University Press.

Slavin, R. E. (1988). Synthesis of research on group-ing in elementary and secondary schools. *Educational Leadership, 46* (1), 67–77.

Slavin, R. E. (1990). Achievement effects of ability grouping in secondary schools: A best-evidence synthesis. *Review of Educational Research, 60,* 471–499.

Sørensen, A. B., & Hallinan, M. T. (1986). Effects of ability grouping on growth in academic achieve-ment. *American Educational Research Journal, 23,* 519–542.

Voigt, J. (1994). Negotiation of mathematical mean-ing and learning mathematics. In P. Cobb (Ed.), *Learning mathematics: Constructivist and interac-tionist theories of mathematical development* (pp. 171–194). Dordrecht, The Netherlands: Kluwer.

Wheelock, A. (1992). The case for untracking. *Educational Leadership, 50* (2), 6–10.

Willie, C. V. (1990). Diversity, school improvement, and choice: Research agenda items for the 1990s. *Education and Urban Society, 23* (1), 73–79.

Wood, T., & Yackel, E. (1990). The development of collaborative dialogue within small group inter-actions. In L. P. Steffe & T. Wood (Eds.), *Transforming children's mathematics education: International perspectives* (pp. 244–252). Hillsdale, NJ: Erlbaum.

Wortman, P. M., & Bryant, F. B. (1985). School deseg-regation and Black achievement. *Sociological Methods & Research, 13,* 289–324.

Yackel, E., Cobb, P., & Wood, T. (1991). Small-group interactions as a source of learning opportunities in second-grade mathematics. *Journal for Research in Mathematics Education, 22,* 390–408.

APPENDIX

Calculation of Effects

The effects of grouping and the initial differences on achievement in each school were calculated separately for each posttest. The overall grouping effect α_H was defined as equal to $\sum_{j=1}^{m-1} H_j$, in which m indicates the number of ability groups in the school and H_j is the effect of ability group j (see Figure 4). Similarly, the overall effect α_P of the initial differences between the students was defined as equal to $\sum_{j=1}^{m-1} P_j$, in which P_j is the effect of initial differences for ability group j (see Figure 4). The overall difference D (in *SD* units) between the strongest and the weakest students for each posttest is $D = \alpha_P - \alpha_H$. The following two examples are taken from Table 7.1:

1. In School 1, $\alpha P = 2.75$, $\alpha H = -0.36$; therefore $D = 2.75 - (-0.36) = 3.11$. The latter result means that the actual difference between the strongest and the weakest students (3.11) was greater than the expected one (2.75). The added gap was 0.36.

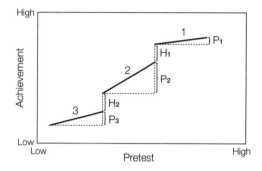

Figure 7.4. A hypothetical example of initial differences ($P_1 + P_2 + P_3$) and grouping effect ($H_2 + H_1$).

2. In School 11, $\alpha_P = 2.20$, $\alpha_H = 0.26$; therefore $D = 2.20 - 0.26 = 1.94$. The latter result means that the actual difference between the strongest and the weakest students (1.94) was smaller than the expected one (2.20). The reduced gap was 0.26.

The actual calculation of effects used the multiple regression equation of posttest scores on pretest scores and ability-group levels. In this design the regression coefficients of ability level and pretest equal the mean, across ability groups, of the effects of ability level and initial differences, respectively. Thus, the overall effect of grouping in each school is $(m - 1) \beta_H$, in which β_H is the regression coefficient of ability level. The overall effect of initial differences is $(X_{max} - X_{min}) \beta_P$, in which β_P is the regression coefficient of the pretest score and $(X_{max} - X_{min})$ is the range of the pretest scores within the school (Cook & Campbell, 1979).

MIXED-ABILITY VERSUS SAME-ABILITY GROUPING IN MATHEMATICS

Liora Linchevski, The Hebrew University of Jerusalem
Bilha Kutscher, The David Yellin Teachers College

Abstract. Findings from 2 studies of the effects of learning mathematics in mixed-ability settings compared with the effects of learning in same-ability settings indicate that the achievements of students need not be compromised in a heterogeneous setting. On the contrary, the achievements of the average and less able students proved to be significantly higher when compared to those of their peers in the same-ability classes, whereas no significant difference was observed between the performances of the high-ability students who learned in the two settings.

THE degree of influence of school-grouping methods on the individual student's scholastic achievement is a central issue in educational research. One of the most widespread methods of grouping students in the same grade is ability grouping, either on a subject-by-subject basis (*tracking*) or for all subjects at once (*streaming*). Tracking and streaming are widely viewed as the best ways to improve the scholastic achievements of all students. Ability grouping has been justified on the basis of the need to adapt class content, pace, and teaching methods to students who are functioning on different levels. In mathematics, it is also justified by the *nature* of the subject. Mathematics is perceived as graded, linear, structured, serial, and cumulative—making work with groups of students having different levels of knowledge and ability difficult.

This article is adapted from Linchevski, L., & Kutscher, B. (1998). Tell me with whom you're learning, and I'll tell you how much you've learned: Mixed-ability versus same-ability grouping in mathematics. *Journal for Research in Mathematics Education, 29,* 533–554.

Recently, however, doubt has been cast on whether placing students into ability groups is the correct method for dealing with diverse abilities. Research has generally shown that the scholastic achievements of students assigned to higher tracks are better than those of students who are judged to have similar abilities but who have been placed in lower tracks. Researchers conducting studies of this sort have concluded that the placement of students in ability groups in and of itself increases the gaps among students beyond what would be expected on the basis of the initial differences among them. In fact, before our two studies began, one of us was involved in a study (Cahan & Linchevski, 1996) of the effect of tracking in mathematics in junior high schools. The findings of that study clearly indicated that the differences among the scholastic achievements of the students at the different group levels at the end of the first and the third years of junior high school were greater than would be predicted by the placement scores. Moreover, in most of the schools, after 3 years, the effect of the group level was greater than the effect of the initial differences among the students.

What has not been shown, however, is (a) whether this growth in inequality is avoided in mixed-ability settings and (b) whether the achievement gap that results from tracking occurs because tracking helps students in the higher ability groups, because tracking harms students in the lower ability groups, or because of some combination of the two factors. Perhaps the widening gap between stronger and weaker students occurs in heterogeneous settings as well as in homogeneous settings. The two studies described here were designed to address these issues. The studies took place as part of a large project in Israeli junior high schools in which mathematics is taught in mixed-ability settings.

STUDY 1

Study 1 was designed to answer the question "Is the gap between better and weaker students learning together in mixed-ability settings for a certain period of time different from the gap that would be expected on the basis of initial differences between the two groups?" To investigate this question, we used the same research design used in the earlier Cahan and Linchevski (1996) study, in which the gap between better and weaker students increased in same-ability classes. The use of identical research designs made possible the comparison between the studies.

Design

For this study, the unit of analysis was a school; that is, we compared scores among schools rather than among classes or students. Because teaching in the participating schools actually took place in heterogeneous settings, we undertook a hypothetical division of the students into the various group levels *as if* these classes were actually going to study in separate ability groups. At the beginning of the seventh grade, each school therefore assigned each student to an ability-group level according to the school's previous tracking policy, although in reality these students studied mathematics in heterogeneous settings. This assigning procedure was done without the knowledge of the students' mathematics teachers. The hypothetical group level and the placement scores (the pretest) were compared with the achievement test scores in mathematics after 1 year and after 2 years—at the end of the seventh grade and at the end of the eighth grade. The effects of grouping and initial differences on achievement were calculated separately for each school.

Seventh-grade students in the 12 Israeli junior high schools that participated in the project were, in addition to being placed hypothetically on the basis of their entrance data, tested at the end of the seventh grade. We have complete data for 1629 of these students. In 4 of the 12 schools, the (389) students were also tested at the end of eighth grade.

Students' achievements in mathematics were measured by tests constructed according to the topics covered in the schools, as detailed in the Israeli national mathematics curriculum (first- and second-year algebra, problem solving, and geometry). The tests were validated by experts and by the General Inspector for Mathematics Teaching in the Israeli Ministry of Education. The test items were open ended, allowing for more flexible types of questions to be asked. To control for between-school differences in mean achievement levels and test-score variance, the scores on pretests and posttests were standardized separately for each school.

Results

The achievement scores in 10 of the 12 schools at the end of seventh grade and in 3 of 4 schools at the end of eighth grade showed that the hypothetical grouping at the beginning of seventh grade, when examined as it related to the initial differences found in the pretest, were not statistically significant in these schools; that is, the gap between high-ability and low-ability students did not increase. This study was designed to answer the question "Is the gap between better and weaker students learning together in mixed-ability settings for a certain period of time different from the gap that would be expected on the basis of initial differences between the two groups?" The answer to this question on the basis of the research results is definitely no. Generally speaking, we concluded that in heterogeneous classrooms the differences in achievement are explained mainly by the initial differences found at the beginning of seventh grade. We did not find the widening gap between better and weaker students that was found in our earlier study using same-ability classes. Thus, the main difference between the results of this study and those of the same-ability-group study previously described (Cahan & Linchevski, 1996) is the near absence of a group effect in this mixed-ability-group study and the presence of a grouping effect in all the schools in the previous same-ability-group study. That is, in this study, mixed-ability grouping did not widen the gap between better and weaker students, whereas in an earlier study, same-ability grouping did result in widening the gap.

STUDY 2

Next we compared the mathematical achievements (actual grades) of students placed in same-ability classes with those of students placed in mixed-ability classes within the same school. The research question investigated was "Which of the two systems—placement in heterogeneous classes as described earlier in the project description or placement in same-ability classes—leads to greater achievement by students?" Moreover, "when parallel levels in each of these systems are compared, which system leads to greater achievement for the better students, the intermediate students, and the weaker students?"

Design

In one Israeli junior high school, unassociated with our project, mathematics was taught in same-ability classes. Although the mathematics faculty considered tracking the only fair, effective way to deal with the heterogeneity of students, they were unhappy with the achievement of students in their lower tracks of mathematics classes. After a long process of deliberation that included meetings, discussions, and reading some of the relevant research literature with our team, the mathematics teacher decided to join our project and to participate in a "real experiment" (for which teachers obtained the parents' agreement). The experiment was designed to test which of the two systems—heterogeneous or homogeneous settings—was more beneficial for the students.

At the beginning of seventh grade, all the students were new to the school; they were randomly assigned to four mixed-ability classes, here called Classes A, B, C, and D. (All content areas other than mathematics were taught, as in the past, in these homeroom settings.) Then, using the same procedure the school had always used for tracking in mathematics, all students in the four classes were assigned, on paper, to one of three ability-group levels—high, intermediate, or low—for mathematics. However, the information used to track students was applied to only two of the four randomly selected classes. These two groups, Classes A and B, were tracked into three smaller, separate, same-ability mathematics classes according to the assignment procedure, here called H (high), I (intermediate), and L (low). Classes C and D studied mathematics in their original mixed-ability classes. Thus, five mathematics classes were involved in all: two mixed-ability classes (C and D) and three same-ability classes (H, I, and L). In the two mixed-ability classes, the results of the tracking procedures were not disclosed to the teachers or to the students.

Teachers were then randomly assigned to the five mathematics classes. Teachers of both kinds of ability groups were involved in weekly workshops: For teachers of same-ability groups, the school's mathematics coordinator led discussions and presented activities appropriate for students of same-ability levels, and for the mixed-ability-group-teachers' workshops, a member of our research group led discussions and presented activities appropriate for the heterogeneous classes.

The students remained in the five groups for 2 years. At the end of eighth grade, achievement tests were administered to all the students. Two alternatives had been discussed with the teachers: One was to give all students the same test, regardless of their placement; the other was to write a test for each ability-group level, with students in the true same-ability classes and students in the equivalent hypothetical level in the mixed-ability classes being given the same test. The teachers thought that the second alternative was a way to reduce anxiety of students who had learned in the lower same-ability classes and had been accustomed to tests especially prepared for their level. Finally, we decided that both forms of testing would be used. The questions for these tests were proposed by the teachers and the mathematics coordinator, and the final versions were written by a representative of the Ministry of Education.

Results

Table 8.1 displays the average final-achievement-test scores for each group level for the ability-group tests and for the common test given to all students. The data have been presented for same-ability groups and for mixed-ability groups. On both test versions, the average scores of high-level students in the same-ability classes were higher (but not significantly so) than those of the students in the mixed-ability classes. Also on both versions, the average scores of intermediate- and low-level students in the same-ability grouping system were significantly lower than those in the mixed-ability system. Moreover, most low-level students in the same-ability classes seemed to be unable to answer the questions on the common test, inasmuch as they handed in almost empty test papers, whereas the corresponding students in the heterogeneous classes scored an average of 54%. The students in the mixed-ability classes who had been hypothetically assigned to the intermediate and low tracks found the differential tests, written for the actual lower tracks, relatively easy; they were accustomed to much higher demands and expectations.

Table 8.1

Achievements (in Percentages) in Mathematics at the End of Eighth Grade

Tests	Same-ability groups			Mixed-ability groups		
	High	Intermediate	Low	High	Intermediate	Low
Differential ability test [a]						
Mean	85	64	55	82	80*	78*
SD	7.8	5.6	6.2	7.5	4.3	5.1
N	33	27	14	35	26	15
Common test						
Mean	88	41	—[b]	85	65*	54
SD	8.1	5.1	—	6.9	6.1	3.9
N	33	27	14	35	26	15

Note. * indicates a statistically significant difference between the same-ability mean and the mixed-ability mean.

[a] The differential ability test was actually three tests, given to students in the appropriate same-ability classes and to the students who would have been in those same-ability classes had not they been placed in mixed-ability settings for the purposes of this study. That is, the students in the mixed-ability classes were all assigned to take high-, intermediate-, and low-group tests only on the basis of the pretest. All had received the same, untracked instruction.

[b] Because many of these students did not complete the test, we could not statistically compare this group with others, but we felt safe in assuming that a significant difference existed.

An analysis of the test papers showed that the high-ability students in the mixed-ability classes lost points for formal presentation and notation, such as the symbolic notation for a truth set or the formal presentation of geometric proofs, whereas high-level students in the same-ability classes were quite competent in this area. The teachers of the mixed-ability classes confirmed that some symbolic notations and formal presentations that traditionally constitute a major problem for average students were given less attention in their classes.

CONCLUSIONS AND DISCUSSION

In these two studies we compared the effects of mixed-ability and same-ability grouping on mathematics performance of students classified as having high ability, intermediate ability, or low ability. In the first study, we found that learning in mixed-ability classes prevented the widening of the gap (usually found when students are grouped by ability) between high- and low-ability students more than would be expected on the basis of the initial differences between the two groups. In the second study, we compared the effects of mixed-ability and same-ability grouping on the mathematics performance of students classified by ability: high, intermediate, or low. The results showed that placement of students into mixed-ability mathematics classes was not detrimental to their achievements when compared to achievements of similar-ability-level students who had learned in separate same-ability classes. On the contrary, the average and weaker students who studied in mixed-ability settings showed significant achievement gains, whereas the loss in achievement of the stronger students was negligible. Demonstrating this result was one of our main goals.

By integrating the results of Studies 1 and 2, we might conclude that an increase in the gap, due to learning in the tracking system, results mainly from the loss in the weaker students' achievements instead of from the stronger students' gains. In Study 1 we have shown, generally, that in mixed-ability classes the gap did not increase nor were achievements significantly impaired. In Study 2 the comparison between the achievements of the mixed-ability students and their same-ability counterparts indicates that the achievements of the average and lower ability groups in the mixed-ability classes were higher. To conclude, our studies indicate that students of all ability levels can learn mathematics effectively in heterogeneous classes.

REFERENCE

Cahan, S., & Linchevski, L. (1996). The cumulative effect of ability grouping on mathematical achievement: A longitudinal perspective. *Studies in Educational Evaluation, 22* (1), 29–40.

Editors' Comments on the Linchevski and Kutscher Chapters

IN THESE comments we describe the changes that we made from the original published research article and explain why the changes were made. We try to then generalize to provide hints you can use to read original research articles and take from those articles whatever you find most relevant.

The original article described three studies. The first two were on ability grouping of students; the third was on teachers' attitudes. We thought that the studies on ability grouping were the most interesting, so we chose not to include the third study. Another reader might, however, be interested in teachers' attitudes and would focus on that study. The reader has a choice of focus, although in this instance we have made the choice for you. Notice that the abstract of the revised chapter was changed to reflect our focus. (The list of key words following the abstract in the original article is provided to help researchers do electronic searches of articles on the *Journal for Research in Mathematics Education [JRME]* Web site.)

The first paragraph of the original article is a good introduction, so we kept that in the revised chapter. The next paragraphs provide the literature review on other research on ability grouping, and we chose to keep only some of the key sentences. Usually teachers are not especially concerned about details on research literature leading up to the present study, and this section can often be skipped in reading research reports. In this case, one previous study, described just before Study 1, provides information that is helpful in interpreting these studies. In this study by Cahan and Linchevski, the only former study we describe, the students were in mathematics classes grouped by ability, and the differences between high-ability and low-ability groups increased over the 3 years of the study. We include this section's last paragraph, in which the authors state the purpose of their studies. The authors of the original article then included a section on the TAP project to provide some background for these and previous studies, but this description is not essential to understanding these studies and so is omitted.

The authors then begin to describe these studies. They provide the rationale for the study and information on its design. Readers can skim this part and return to it if they desire. The rewritten article begins now with Study 1. Only the essential details of the study are included in the condensed chapter. This study was similar to the Cahan and Linchevski study described in the last paragraph, but in Study 1 the researchers worked only with heterogeneous classes.

The Study 1 results provided in the original article are detailed and are based on statistical analysis. A teacher-reader is probably interested not in all the details but rather in the final results, which are stated in the final section of the study. The results of this first study (of learning in mixed-ability groups) and the results of the previous Cahan and Linchevski study (of learning in same-ability groups) are compared. In Study 1 the gap between high-ability students and low-ability students did *not* increase. Notice that neither study included both mixed- and same-ability groups. Why were both groups not studied simultaneously? Unfortunately, researchers usually cannot randomly assign students to mixed-ability or same-ability classes without raising serious objections from teachers, administrators, parents, and students, and thus they resort to studying the learning of students in one situation or the other, as happened in this Study 1 and in the sited Cahan and Linchevski study.

However, in the second study serendipity played a role. Linchevski and Kutscher were approached by teachers at a middle school at which students were grouped by ability because the teachers and administrators believed that arrangement to be the most effective way to teach. Yet they were continually disappointed by the results, particularly for the low-ability groups. They, and the parents of their students, were willing to participate in a study in which the researchers randomly assigned students to mixed- and same-ability groups and compared the achievements of these groups 2 years later. The design of the second study is complex and needs to be read with special care, but we assure you that working through the details will be rewarding.

The results of Study 2 show unequivocally that high-ability students were not harmed by learning in classes of mixed-ability levels but that intermediate-ability and low-ability students were harmed when assigned to classes on the basis of placement-test scores (i.e., tracked). This result is important, particularly because the study was well designed and was carried out over 2 full years. Often, the results section can provide information that is useful to the reader, but some discrimination is required to identify those parts that are most useful.

Finally, we took parts of the Conclusions and Discussion sections of the original article, summarized the information concerning Studies 1 and 2, and kept the important paragraph about integrating the results of the two studies. Once again, we omitted information on Study 3.

These studies were undertaken in Israel. The Israeli school system is in many ways similar to that of the United States. We, the authors and the editors, doubt that the results would be much different if the studies were undertaken in other countries.

We hope that this exercise of comparing an original research article with the edited version has been an interesting one for you. *JRME* presents many articles that are of interest to teachers. *JRME* is available online at the NCTM Web site, which provides new opportunities for one to explore. Although the full articles will not be accessible without a subscription, the abstracts are available, and if an article sounds particularly interesting, you can pay to have it downloaded, or you can seek out the journal in a local university library.

SECTION II

RESEARCH RELATED TO LEARNING: INTRODUCTION

READERS who argue that all the chapters are in some way related to learning are correct. But the studies in this section deal with learning in a very direct way. The authors of these chapters present research results on learning and important constructs that help us think more clearly about learning. Some chapters also include tasks that can be used in instruction or assessment.

Chapter 9. "Learning in an Inquiry-Based Classroom: Fifth Graders' Enumeration of Cubes in 3D Arrays," by Michael T. Battista

Measurement instruction often focuses on students' learning such formulas as $V = lwh$. Yet understanding volume and why this particular formula works for finding the volume of rectangular solids is, for intermediate-grades students, far more elusive than many teachers realize. This carefully designed study took place in a fifth-grade classroom in which the teacher created an environment that encouraged inquiry, problem solving, and sense making. Students, working in pairs, studied one diagram of a box with squares marked along the sides and one diagram in which the box was "flattened" with the sides and base showing. The students explored these diagrams and made conjectures about the number of cubes needed to fill each box. The tasks became increasingly difficult. The role of communication between the students was found to be an important factor in their success on these tasks. The results show that coming to understand the structure of three-dimensional (3D) cube arrays is challenging for fifth graders. Also, the tasks used in this study could be used by teachers who want students to explore ideas of filling space as an avenue to understanding volume.

Lessons learned

From this study the researcher draws important instructional implications that go beyond the content focus of the study. Powerful mathematics learning can occur in problem-centered inquiry-based teaching, but that learning is not necessarily linear. To be successful, students need to grapple with ideas, make mistakes, become confused, and help one another through these difficulties. The teacher's role is not one of intervention so much as one of structuring tasks so that students learn from them.

Chapter 10. "Teacher Appropriation and Student Learning of Geometry Through Design," by Cathy Jacobson and Richard Lehrer

Second-grade students in the classrooms of four experienced teachers used quilt design to learn about transformational geometry. The teachers had participated in a professional development program that focused on teaching number and operation concepts. Two of the teachers had also had professional development to better understand how children learn geometry concepts. Did this additional professional development make a difference in what their students learned? Can good primary teachers effectively teach mathematics concepts if they have not had opportunities to learn how children come to understand those concepts? How specific does a teacher's knowledge of student thinking need to be for that teacher to be an effective teacher of that knowledge?

The authors introduce a construct that they (and others) call *appropriation*. The word is not new, of course—to *appropriate* something is to take it as one's own. To appropriate a curriculum, then, is to take a curriculum as one's own. The authors show, in this chapter, that this process cannot happen unless the teacher becomes so familiar with a curriculum and how students learn the content of that curriculum that it becomes "owned" by the teacher. Furthermore, the researchers found that when teachers appropriate a curriculum, their students learn more than students of teachers who have not, for various reasons, taken ownership of that curriculum.

Lessons learned

A good teacher who has not had the opportunity to study and reflect deeply on a topic and appropriate it as her or his own will not teach that topic as well as a teacher who has had this opportunity. A second lesson learned is that topics can be approached in unique ways, in this example transformation geometry approached though quilting patterns, that provide access to a topic that otherwise might seem too difficult for students at a particular grade level.

Chapter 11. "A Longitudinal Study of Invention and Understanding: Children's Multidigit Addition and Subtraction," by Victoria R. Jacobs

Most children learn the standard algorithms (procedures) for addition and subtraction in the primary grades without exploring any other options. But some teachers encourage students to invent procedures for adding and subtracting. For example, a student asked to solve 62 – 28 might say, "Sixty-two take away 20 is 42. Then take away 8: Take away 2 is 40; take away 6 more is 34." In this study, students were encouraged to solve problems in any way they could; the children typically used invented procedures, but they also learned standard procedures. The researchers followed students over their first three grades to explore the effects of allowing students to invent their own strategies. They found, and the author discusses here, several advantages for allowing students this flexibility in learning to add and subtract numbers. They argue not that invented strategies are better than standard algorithms but rather that those stu-

dents who invent strategies learn much more than simply how to add and subtract; they acquire a deeper understanding of place value and are better problem solvers.

Lessons learned

At all grade levels, we teachers teach standard procedures for carrying out mathematical operations. But some students seem never to really learn them. Could it be that they need to spend more time exploring and coming to understand what the procedures are intended to do for them? Teachers who respect the development of conceptual understanding are more likely than others to foster in their students a deep understanding of the content being taught. A second lesson to take from this study is that learning happens over years, so that long-term studies can provide information that cannot be found from short-term studies.

Chapter 12. "Interference of Instrumental Instruction in Subsequent Relational Learning," by Dolores D. Pesek and David Kirshner

Sometimes we all contend with phenomena we recognize but cannot quite "pin down." Having someone put a name to such phenomena helps us think more clearly about them, as seems to be true with Skemp's descriptions, introduced in the 1970s, of *instrumental* understanding and *relational* understanding. Students who learn instrumentally are those who learn rules only, but those who learn relationally learn not only rules and procedures but also when and why to use them. These constructs have gained universal acceptance and appear in mathematics education literature from many countries. Giving names to these phenomena helps people converse about them with the knowledge that others understand what they mean.

Ideally we want our students to gain relational understanding of what we teach them, but as Pesek and Kirshner recognize, teachers who want their students to learn relationally are often in a difficult position. On the one hand, professional organizations, researchers, and professional developers have convinced them that students should understand the mathematics they learn; that is, students should learn relationally. On the other hand, because of the accountability movement and high-stakes testing, they believe that they must teach their students to pass tests that often call only for instrumental learning. What is the teacher to do?

What are the effects of teaching for instrumental understanding part of the time and teaching for relational understanding part of the time? These researchers explore this difficult conundrum. They demonstrate that students who learn procedures are unwilling and perhaps unable to later come to understand the reasons that the procedures work.

Lessons learned

Teachers are caught in the middle of this difficult question of whether to teach for skills or for understanding. Can both be done? Can students who learn only relationally be able to compete, in the realm of testing, with students who learn instrumentally, with perhaps a relationally oriented lesson tacked on at the end? The answers provided here are not completely satisfactory, because research studies such as this one are limited in scale. Yet the results help teachers better understand the difference between, and possible effects of, teaching relationally or teaching instrumentally. Teachers can, on their own, profitably undertake some classroom research in this area to help them arrive at satisfactory answers they can then share with administrators and parents.

Chapter 13. "An Exploration of Aspects of Language Proficiency and Algebra Learning," by Mollie MacGregor and Elizabeth Price

Language proficiency is usually thought of in terms of English proficiency (or proficiency in the official languages of a country). But even English speakers may not have the linguistic proficiency necessary to be academically successful. These authors explore the linguistic requirements necessary for successful algebra learning. They argue that metalinguistic awareness in ordinary language has a counterpart in algebraic language. (*Metalinguistic* refers to the interrelationships between language and other cultural behavior.) All teachers of algebra have students who misinterpret algebraic symbols and syntax. The researchers examine these factors, together with what they call *awareness of potential ambiguity,* as they affect algebra learning. Ambiguity exists in recognizing, for example, that an expression can have more than one interpretation.

This research area is new, and the authors are charting new territory in their exploration for reasons some students have difficulty learning algebra. In this study, they found a positive correlation between language and algebra scores. This chapter not only contains some interesting research findings but also introduces new constructs, such as awareness of potential ambiguity, and interesting test items that have been piloted and used with many students.

Lessons learned

Language proficiency can affect mathematics learning, particularly algebra learning, in many ways. In many states in the United States, all eighth graders are required to take algebra, and evidence shows that many fail. Do students fail because they are not mathematically prepared, or is linguistic proficiency also a factor? Using the constructs and test items provided here, teachers may want to try some research in their own algebra classes.

Chapter 14. "Developing Concepts of Sampling for Statistical Literacy," by Jane M. Watson and Jonathan B. Moritz

The study of probability and statistics now begins in the early grades, at least in an introductory fashion. Children sometimes undertake small-scale statistical studies in their classrooms. How do they go about selecting a sample from which to draw data? The concept of an unbiased sample takes many years to develop. In this study, the researchers interviewed students in Grades 3, 6, and 9 and then categorized the characteristics of students' construction, over time, of the concept of sample. The progression of understanding through these grades is quite remarkable. Most of the third graders viewed a sample as a small amount of something, like a taste of something handed out in a supermarket. By sixth grade, the students showed some understanding of the fact that a sample should indicate something about the "whole," that it is in some way an "average" and that it should not be biased, but they were often unable to recognize bias. By ninth grade, most of the students clearly understood that the sample had to represent the population without bias.

Lessons learned

Curriculum writers will find this study helpful in determining how to gauge when students are ready to move to the next level of work on sampling. But teachers too will find this study helpful in determining where their students stand with respect to development of the concept of sample. The items used with students in the study are included, as are students' typical responses at various levels.

Chapter 15. "When a Student Perpetually Struggles," by Kristine K. Montis

A fifth-grade student, Kay, struggled with little success throughout her learning of mathematics. While tutoring Kay, the author came to better understand how Kay's phonological-processing deficits affected her mathematics learning. These deficits are not detected in standard hearing tests; a student can hear the sounds but cannot accurately process them. After identifying Kay's problem, the researcher was able to make progress in helping Kay, particularly in the area of fraction understanding. Kay did not distinguish between *eight* and *eighths*, for example. She saw the numerator and denominator as unconnected separate entities. The author refers to Kay's "fuzzy perception of speech sounds," and when Kay was helped to make the necessary distinctions, she began to make sense of fractions. Several indications of this particular deficit are listed in the chapter. The close connection between language learning and mathematical learning is also noted.

Lessons learned

We all have had Kays in our classes—the quiet, hard-working students who never quite "got" it. Even being aware of the role of phonological processing in learning mathematics may help some teachers reach their own Kays. The author points out that Kay did not fit the legal definition of *learning disabled* but had been diagnosed as a *slow learner* who needed only more time rather than special approaches or materials. However, in Kay's case, a special approach and use of materials helped her learn.

LEARNING IN AN INQUIRY-BASED CLASSROOM: FIFTH GRADERS' ENUMERATION OF CUBES IN 3D ARRAYS

Michael T. Battista, Kent State University

Abstract. This study illustrates how powerful mathematics learning can occur in problem-centered inquiry-based teaching. Given properly designed instructional tasks and an appropriate classroom culture of inquiry, students constructed, revised, and refined their concepts and theories to make sense of and solve the problem of enumerating cubes in rectangular boxes. Difficulties students faced and the ways they overcame those difficulties are illustrated.

To DEVELOP powerful mathematical thinking, instruction must carefully guide and support students' personal construction of concepts and ways of reasoning while the students intentionally try to make sense of situations. To be effective, this guidance and support must be "grounded in detailed analyses of children's mathematical experiences and the processes by which [children] construct mathematical knowledge" (Cobb, Wood, & Yackel, 1990, p. 130). In this chapter I discuss the initial concepts and reasoning that students must construct in developing a genuine understanding of volume.

3D ARRAYS OF CUBES

Research has shown that students have considerable difficulty in determining the number of cubes

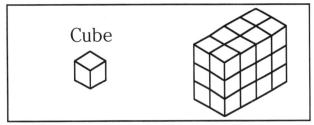

Figure 9.1. The rectangular building consists of cubes.

contained in 3D rectangular buildings like the one shown in Figure 9.1.

For example, Ben-Chaim, Lappan, and Houang (1985) found that fewer than 50% of middle-grades students could solve such problems, with about 25% of fifth graders, 40% to 45% of sixth and seventh graders, and 50% of eighth graders answering correctly. The results of the second NAEP showed that fewer than 40% of 17-year-olds solved problems of this type (Hirstein, 1981). Recent research indicates that because their spatial structuring of the array is incorrect, many students are unable to correctly enumerate the cubes in such an array (Battista & Clements, 1996). *Spatial structuring* is the mental act of constructing an organization or form for an object or set of objects; through it one identifies an object's spatial components and establishes interrelationships among those components.

THE INSTRUCTIONAL SETTING

The fifth-grade students observed in this study were taught by a teacher who was highly skilled in creating a classroom culture of inquiry, problem solving, and sense making. In early February, the teacher taught a 4-week instructional unit on volume to her two regular mathematics classes, which

This chapter is adapted from Battista, M. T. (1999). Fifth graders' enumeration of cubes in 3D arrays: Conceptual progress in an inquiry-based classroom. *Journal for Research in Mathematics Education, 30,* 417–448.

Support was provided by NSF Grant RED 8954664. The opinions expressed do not necessarily reflect the views of the Foundation.

How Many Cubes?

How many cubes fit in each box? <u>Predict,</u> then build to check.
Check your prediction for a box before going on to the next box.

	Pattern Picture	*Box Picture*
Box A		
Box B		
Box C		
Box D		
Box E		
Box F	The bottom of the box is four units by five units. The box is three units high.	

Figure 9.2.

How Many Cubes? activity (reduced in size).

each met for 1 hour per day. The teacher explained to the students that their goal for the first activity (see Figure 9.2) was to find a way to correctly predict the number of cubes that would fill boxes described by pictures, patterns (nets), or words.

The students worked collaboratively in pairs, first predicting the number of cubes that would fit in a graphically represented box, then checking

their answer(s) by making the box from grid paper and filling it with cubes (the students checked their results for one problem before going on to the next). The teacher circulated about the room, interacting with student pairs, encouraging within-pair communication and collaboration, and promoting individual sense making. Student-pair work on this activity was completed after 2 1/2 class periods. (In this report I discuss only one of the three student pairs examined in the original study.)

THE EVOLUTION OF STUDENTS' THINKING DURING INSTRUCTION

Episode 1, Day 1

For Box A (see Figure 9.2), N and P silently analyze the pictures, then explain their thinking to each other. N counts the 12 outermost squares on the 4 side flaps of the pattern picture (see Figure 9.3a), then multiplies by 2: "There's 2 little squares going up on each side, so you times them." P counts the 12 visible cube faces on Box Picture A, then doubles that number for the hidden lateral faces of the box. The boys agree on 24 as their prediction.

> *P:* [After putting 4 rows of 4 cubes into the box] We're wrong. It's 4 sets of 4 equals 16.
>
> *N:* What are we doing wrong? [Neither student has an answer, so they move on to Box B.]
>
> *P:* What do you think we should do? [Pause.] [Pointing to 2 visible faces of the cube at the bottom right-front corner of Box Picture B] This is 1 box [cube], those, 2.
>
> *N:* Oh, I know what we did wrong!
>
> *P:* Is that what we did, at the corners?
>
> *N:* Yeah, we counted this [pointing to the front face of the bottom right-front cube]

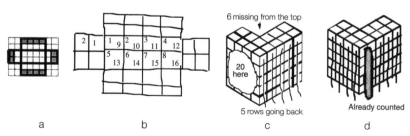

Figure 9.3. Work of N and P.

and then the side over there [pointing to the right face of that cube].

P: So we'll have to take away 4 [pointing to the 4 vertical edges of Box Picture A]; no, wait, we have to take away 8. [P subtracts 8 from their prediction of 24 and tells N that this subtraction would have made their prediction correct.]

[In their prediction for Box B, P counts 21 visible cube faces on the box picture, then doubles that number for the box's hidden lateral faces. He subtracts 8 for double-counting (not taking into account that this box is 3 cubes high, not 2, like Box A), predicting 42 – 8 = 34. He asks N whether this prediction makes sense. N hesitates, then makes his own prediction. He adds 12 and 12 for the right and left lateral sides of Box Picture B, then adds 3 and 3 for the middle column of both the front and back, explaining that the outer columns of 3 on the front and back were counted when he enumerated the right and left faces. He predicts 30.]

Commentary

For Box A, N and P made predictions on the basis of different mental models for the cube array: N looked at the pattern picture and multiplied by 2 for the "two squares going up." P looked at the box picture and multiplied by 2 for the two lateral sides that could not be seen in the box picture. Although the boys agreed on a prediction of 24, the discrepancy between their predicted and actual answers caused them to reflect on and reevaluate their enumeration strategies.

When P applied his enumeration strategy to Box Picture B, he focused on the two visible faces of the bottom-right corner cube. By coordinating the positions of these faces in 3D space, he was able to realize that they were, in fact, the front and right faces of the same cube. When N and P extended this new interpretation of adjacent faces to other vertical-edge cubes, they recognized that P's current enumeration strategy double-counted such cubes. The attempts the boys made to correct their enumeration strategies, however, dealt with the double-counting error in different ways. P compensated for the error by subtracting the number of cubes he thought he had double-counted, whereas, by spatially restructuring the array, N attempted to

imagine the cubes so he would not double-count them.

Episode 2

After Box B is constructed, P puts several 1-by-1-by-4 rows of cubes into it.

P: There's 4 times 3; that's 12 [counting cubes he placed in the first layer]. [After filling the middle layer with 2 rows of 4 cubes] Then there'll be 12 again in the middle—24.

[However, P stops enumerating and waits until they fill the last layer. The boys then count all the cubes by ones while they take the 1-by-1-by-4 rows out of the box, getting a total of 36. They silently reflect for a minute or two. P starts putting the rows of cubes back into the box, but N stops him when he has the bottom layer filled. N removes the cubes and examines the box, seeming to count the squares on various sections.]

P: What was your answer, 30?

N: Yeah [looking puzzled].

[P puts the rows back into the box, then takes them out again, and counts the cubes "just to make sure," again getting 36. N is staring off into space, deep in thought.]

P: What are you thinking we should do?

N: Well, I was thinking I forgot some in the middle, 3 in the middle.

P: I got it! We were doing minus 8 [pointing to Box Picture A], but really there are 1, 2, 3—4, 5, 6—7, 8, 9—10, 11, 12 [pointing successively to the 4 vertical edges of Box Picture B]. [He looks at Box Picture B and does some recounting but seems puzzled, possibly because subtracting 12 instead of 8 from 42 still does not give the correct answer.] If there's 21 here and 21 there [referring first to the front and right sides, then to the back and left sides], there's still some left in the middle. I got it. We missed 2 in the middle. There's 2 more blocks in the middle, and that would make 36. [Recall that P had predicted 34.]

N: Yeah.

Commentary

The discrepancy between N and P's predicted and actual answers for Box B led them to reflect further on the situation. The boys' reflections, however, were determined by their *current* mental models and enumeration strategies. N focused on structuring the arrays so that he could correctly locate and count the cubes—he seemed intent on determining where the cubes were located. Reflecting on his spatial structuring caused him to recognize that he had omitted "some [cubes] in the middle." But he could not yet structure the cube array to imagine all these cubes—he thought that he had missed 3 when, in fact, he had missed 6. He could not properly coordinate the right-side orthogonal view with the top or front views to create a correct mental model of the interior.

At first, P thought that his prediction error arose from his failing to account for the building height in his strategy of enumerating all outside cube faces and then compensating for double-counting vertical-edge cubes. However, when P's attempted adjustment for this error did not give the correct answer, he reflected on N's comment about failing to count cubes in the interior of the array, deciding that the prediction of 34 missed 2 cubes in the middle. Note that although P reflected on his mental models of the cube array and box picture (thinking spatially), his final enumeration ultimately seems to have been determined by the fact that 2 was the difference between his prediction of 34 and the actual answer of 36 (a strictly numerical analysis).

Although neither P nor N had yet developed a spatial structuring of 3D arrays that led to correct enumeration of cubes, they were making progress. While they reflected on, analyzed, discussed, and revised their work, they abstracted important aspects of the spatial organization of the arrays; these abstractions would later help them make the needed restructuring. Also, even though the boys' enumeration strategies had now diverged, they were still listening to, and attempting to make sense of, each other's comments.

Episode 3

N and P jointly count 21 outside cube faces for Box C (see Figure 9.2), not double-counting cubes

on the right-front vertical edge. They then multiply by 2 for the hidden lateral sides and add 2 (which is the number of cubes that they concluded they had missed in the middle of Box B) for the middle cubes. N thinks that more than 2 might be in the middle but does not say how many. Their prediction is 44 cubes. While N draws the pattern for Box C on grid paper, P continues to think about their prediction.

P: I think I know what we did wrong with this one [Box B]. We counted these [vertical edges] again.

N: But then our prediction would have been more than we got [but it was not].

P: Oh, yeah. Don't we have to take away 12 from 44 [for C]?

N: Why?

P: Just count it again.

[P counts 24 visible cubes on Box Picture C, then multiplies by 2 to get 48. He subtracts 12 for double-counting the vertical-edge cubes, getting a total of 36. This number of cubes is the same as the number they found for Box B, and because they decide that Box C is bigger than Box B, they reject this answer. So P decides that they need to add 2 for the middle, getting 38, but he cannot convince N that this prediction is correct. The boys make and fill the box and find that it contains 48 cubes. They are puzzled, but class ends before they have time to discuss the problem further. On Day 2, N silently points to, and counts the squares in, the center portion of Pattern Picture C three successive times, then asks whether, in their original prediction, they had gotten 42 and then added 2 for the middle.]

Researcher: You counted 21 here [visible cube faces on Box Picture C], times 2, plus 2 for the middle, makes 44.

P: Oh yeah, we did plus 2, but we had to do these 2 right there [pointing to faces along the front-left vertical edge and the back-right vertical edge]. We only did these 2 here [pointing to faces along the front-right vertical edge and the back-left vertical edge]. So we should've gone 6.

Commentary

At the end of Day 1, while P discussed their erroneous prediction for Box B, N listened and pointed out flaws in P's reasoning. Also, although the boys need a joint strategy for their prediction for Box C, P later rejected it and reverted to his original strategy. The boys then decided, by comparing Boxes B and C, that P's subsequent prediction of 48 was unreasonable. P then altered his prediction by attempting to account for interior cubes.

On Day 2, N and P reflected on their incorrect predictions for Box C. P concluded that they had failed to count some of the cubes in the four vertical edges. However, as in his previous adjustments, P derived the number to be added to compensate for these omitted cubes by comparing their predicted and actual answers, not by finding an error in his spatial structuring. (He concluded that when they added 2, their answer was 4 less than 48, so they should have added 6 instead.) Not only was P's spatial structuring of the cube array inadequate, but also his numerical reasoning was not properly linked to that structuring.

Episode 4

> *N:* I think I know Box D; I think it's going to be 30; 5 plus 5 plus 5 [pointing to the columns in the pattern's middle], 15. And it's 2 high. Then you need to do 3 more rows of that because you need to do the top; 20, 25, 30 [pointing to middle columns again].
>
> *P:* I don't know—that'd probably work, I guess.

[After the boys make a box and find that it contains 30 cubes as N predicted, both boys smile. P then asks N to explain his strategy again; N does.]

> *P:* So there's 15 right here [pointing to the bottom of the box].
>
> *N:* Yeah, on the bottom. And in the top part here [motioning], there'd be another 15.
>
> *P:* All this top here [motioning along the top rows of the sides of the box]?
>
> *N:* Yeah. See, you go 1, 2, 3, . . . , 15 [counting the squares on the bottom of the box].

> *P:* And then you add another 15.

Commentary

In Episode 4, N made a correct prediction by structuring the array into 2 layers of 3 columns of 5 cubes. Because he had suspected that a better way existed to solve the problem, he had reflected on and analyzed the situation, coming up with a new strategy. Only after P saw that N's strategy gave the correct answer did P become interested in really listening to, and making sense of, N's strategy description.

Episode 5

On the next problem [Box Picture E], because neither boy is able to employ their new layering strategy in this different graphic context, both return to variants of their old strategies. After the boys complete the pattern for Box E, N applies his layering strategy, predicting "16 times 2," but he is not yet confident in its validity. When the boys find that 32 cubes actually fit in the box, however, N comments, "It is 32," seemingly realizing that his new layering strategy is, in fact, valid in this situation.

Episode 6

The boys are unwilling to make a prediction for Box F until they draw the pattern. Once the pattern is drawn, N silently points to and counts the squares in its middle section, 1–20 for the first layer, 21–40 for the second, and 41–60 for the third.

> *P:* You counted 3 times, actually 4.
>
> *N:* [With assurance] Why 4? It's 3 up.
>
> *P:* Oh yeah, that's right.

[The boys build the box, fill it with rows of 5 cubes, then count the cubes by fives to 60. They do not seem *relieved* that their count matches their prediction but, instead, *expect* their answer to be correct. N and P then try to describe their method of enumeration.]

> *N:* You count the bottom ones, then how many times there is up. Like, if there's 5 on the bottom [pointing to a column of 5 in Box Pattern D] and 3 rows [pointing to a row of 3] and 2 going up [motioning on the

pattern], then you add 15 and 15 because there'd be 3 rows of 5 and 2 rows going up, and you'd get the answer of 30. [Then explaining the pattern diagram he has drawn in his journal (see Figure 9.3b)] You count 1, 2, 3, . . ., 8 [pointing to squares in the middle rectangle of the pattern]. Then it goes 2 up [pointing to the written 1, 2 on the left flap]. So you've counted the bottom row; then go 9, 10, 11, . . ., 16 [points to the same middle squares].

P: You count how many are on the bottom. Then you times that by how many are going up [pointing to a pattern picture].

[To probe P's understanding, the researcher asks how many cubes would fit in a box that is 3 by 2 on the bottom and 5 high. P correctly draws the box's pattern on graph paper.]

P: [It would hold] 3 times 2 on the bottom, 6; 6 times 5 equals 30.

Commentary

Once the boys had drawn the pattern for Box F, N returned to his layering strategy, quite confident not only that he understood the location of the cubes but also that his newly developed strategy gave a correct count. Although N performed the enumeration, P seemed to follow what N was saying, enough, at least, to think that the prediction of 60 was valid, so he did not have to make his own prediction. By the end of the episode, both boys had sufficiently abstracted this strategy so that they could apply it to another box. But notice that even though N led the way in developing a layering enumeration strategy, P's final strategy was more effective because he used multiplication, whereas N used counting or addition.

Throughout their work on the activity sheet, N and P each tried to develop strategies for making correct predictions. Their focus of attention seemed to cycle through predictions; to what actually happened in counting; to attempts to integrate what they abstracted from predicting, counting, and discussing. Discrepancies between their predicted and actual answers caused them to reflect on both their enumeration strategies and their structuring of the cube arrays. Indeed, initially their enumeration strategies were based on more primitive spatial structurings of 3D arrays—they saw the arrays in

terms of the faces of the prism formed. Also, P, especially, seemed to focus more on numerical strategies than on a deep analysis of the spatial organization of the cubes. However, because their initial spatial structuring led to incorrect predictions, the boys refocused their attention on structuring the arrays, with N taking the lead in this restructuring. Finally, while working on Box D, N and P developed the layer-based enumeration strategy that they verified, refined, and became confident in using on subsequent problems.

Episode 7, Day 3

The researcher asks N and P how many cubes would be in a box that has the same bottom as Box A but is 3 cubes high.

P: Eight times 3 equals 24.

N: Yeah, 8, 16, 24. I'm not too good at my multiplication facts.

[The boys then explain how they found the number of cubes in a 3-by-4-by-4 building, given its picture. P made a pattern for the box, found the number of cubes that fit on the bottom (12), then multiplied by 4 high. He got 48. N counted the cube faces on the front-left column of the box picture 4 times (1, 2, 3, 4 going down the column; 5, 6, 7, 8 restarting at the top and going down the column again; and so on, up to 16) then continued this type of counting on the next 2 front columns. He says that he counted each cube 4 times because the box has 4 layers "going back." He comments that counting 3 times would leave off the cubes on the back. The boys then describe their solutions for a second problem, a 4-by-5-by-5 building (see Figure 9.3c):]

P: There's 20 on the left side, times 5 rows going back, equals 100. There's 6 cubes missing from the top, and it's 5 down, so we have to subtract 6 five times: 100 minus 6 equals 94; 94 minus 6 equals 88; 88 minus 6 equals 82; 82 minus 6 equals 76; 76 minus 6 equals 70; 70 is the answer.

N: [Referring to the cube faces in the columns of the left side, as in Figure 9.3d] I counted each of these twice because there's 2 back. I didn't count these [first 2 columns on the right side] because they've already been counted. I then counted each of these twice [referring to the cube faces in the remaining columns of the right side] because they go 2 back this way. [N counted by ones from 1 through 70.]

Commentary

On the second problem, both boys successfully adapted their layer-based enumeration strategies to apply them in the new situation. The boys had developed a powerful and general way of reasoning about an important class of problems. In fact, by the end of the instructional unit, almost all 47 students involved in this instructional unit were capable of properly structuring and enumerating 3D cube arrays.

DISCUSSION

Initially, like N and P, all the case-study students structured their mental models of 3D arrays as uncoordinated sets of orthogonal views. To move beyond this medley-of-viewpoints conception, they had to coordinate orthogonal views; abstract interrelationships among them; and integrate the views into a single, globally coherent, isometric-like mental model (Battista & Clements, 1996). Significantly, the student pairs that in their discussions explicitly dealt with coordination were the most successful in developing appropriate mental models for the cube arrays. For instance, acts of coordination enabled N and P to see that they had double-counted edge cubes and had omitted middle cubes.

Students who developed a layer-structuring that they could correctly apply in predicting the number of cubes in 3D arrays did so by recursively cycling through sequences of acting (structuring and enumerating), reflecting, and abstracting. Their reflection was often comparative in nature, contrasting current with past abstractions. Significantly, students' initial use of a layering structure either in predicting or in enumerating actual cube arrays did not automatically lead to its use in subsequent predictions. Several times, students used a layering structure in enumerating an actual cube array, then, in their next prediction, employed an uncoordinated, locally structured mental model. Why were these students unable to apply this layer structuring in future enumeration attempts? Several factors may be involved.

First, properly structuring 3D cube arrays is a complex process. It results from coordinating and combining mental actions one performs while enumerating and spatially analyzing cube arrays. Because of this complexity, one may need to create a structuring in action several times before it becomes stable enough to be mentally recreated and one may have to recreate and reflect on the structuring several times before it becomes accessible for deeper analysis. Second, if a properly structured mental model were constructed by manipulating actual cube arrays, one might be unable to activate it by seeing a picture of a box or pattern. Or if it were constructed from a box picture, it might not be evoked by a pattern picture. Thus, having students make predictions using a variety of presentation formats (box and pattern pictures, verbal descriptions) may have been essential in engendering students' moves to a sufficiently abstracted spatial structuring that could readily be applied in different contexts.

The power of predictions

Having students first predict, then check their predictions with cubes was an essential component in their establishing the viability of their mental models and enumeration strategies. Because students' predictions were based on their mental models, making predictions encouraged them to reflect on and refine those mental models. Their strategies were refined when the students reorganized their models and strategies to better fit their experiences. Indeed, it is students' mental models that the instructional unit was attempting to develop. Having the students merely make boxes and determine the number of cubes required to fill them would have been unlikely to have promoted nearly as much reflection as having them make and check predictions, because (a) opportunities for reflection arising from discrepancies between predicted and actual answers would have been greatly reduced and (b) students' attention would have been focused on physical activity instead of on their own thinking.

Communication

An individual attempting to make sense of another's communication can interpret that communication only in terms of the conceptual structures he or she can currently activate or quickly construct. In the case-study pairs' collaborations, the students constantly had to refocus their attention and build appropriate mental models to make sense of their partners' communications. But because this refocusing and building took place while the students were struggling to develop their own properly structured mental models of the cube arrays, the students were especially vulnerable to miscommunication. Without sufficient effort to ensure proper communication—taking special care to point out to their partners exactly to what they were referring and attempting to monitor their partners' comprehension—effective communication broke down, causing students either to misunderstand what was being communicated or to "get lost." This unproductive communication, in turn, distracted students from their efforts to develop appropriate mental models for the arrays.

In fact, effective communication was a key factor that distinguished the collaboration of successful and unsuccessful student pairs. For example, N's and P's communications about double-counting and omitting middle cubes were effective enough to cause the boys to restructure their mental models of the arrays and make corresponding changes in their enumeration strategies. And when N finally devised a viable prediction strategy, P was able to make sense of it. The situation was very different for another, less successful case-study pair. The students in this latter pair never seemed to explicitly deal with miscommunications, nor did they address the fact that they had created different structurings and mental models for the cube arrays.

Students in successful pairs also seemed to try harder than those in less successful pairs to understand their partners' comments and to resolve disagreements; they had less tolerance for discrepancies. Indeed, tolerance for discrepancies—both within one's own conceptions and between one's own and those of others—impedes sense making and the development of coherent theories because it suppresses attempts at removing conceptual inconsistencies.

CONCLUSIONS AND INSTRUCTIONAL IMPLICATIONS

This study illustrates how powerful mathematics learning can occur in problem-centered inquiry-based instruction. In an appropriate classroom culture of inquiry, students, like scientists, construct, revise, and refine theories to solve and make sense of perceived problems. Unlike scientists' explorations of uncharted intellectual territories, however, students' theory building is guided by teachers and instructional materials. During the instructional period for the case studies, guidance of students' theory building came mainly through the sequence of problems on the activity sheet. Within the culture of inquiry that had been established in this classroom, the gradual fading of perceptual scaffolding in the sequence of problems, along with the combination of making predictions and then checking those predictions with cubes, induced students to construct increasingly sophisticated structurings and enumeration procedures. But direct teacher intervention is sometimes critical to students' progress. In fact, this teacher had to recognize and explicitly deal with several students' lack of a proper mental model for cube arrays. She had to find ways to get those students to construct more appropriate spatial structurings of the arrays.

This study also illustrates that in problem-centered inquiry-based learning—as in the scientific endeavor in general—the development of viable theories and concepts is often confused, irregular, and untidy. Students, like scientists, make mistakes and struggle to make sense of ideas. Consequently, teachers attempting to implement problem-centered inquiry-based instruction must understand that, often, grappling with ideas and even becoming confused (within reason) are natural components of students' sense making. This situation is unlike that of traditional instruction, in which students' struggles are generally interpreted as deficiencies in the students, the curriculum, or the teacher. Indeed, one of the most difficult things for teachers to learn in implementing problem-centered inquiry-based instruction is when to intervene with students and when to let them struggle. Only by thoroughly understanding the usual paths students take in learning particular mathematical ideas—including stumbling blocks and learning plateaus in these paths—can teachers know when to intervene.

Finally, this research has several implications for understanding effective mathematics instruction. First, to design effective instructional sequences, one must understand the precise constructive itineraries that students follow in learning particular mathematical ideas. Second, properly structured problem-based-inquiry teaching can be very effective in engendering meaningful mathematics learning in students. However, the effectiveness of this approach depends not only on students' having repeated and varied opportunities to construct and employ the targeted concepts but also on their abilities to effectively and independently check the viability of their developing mental models. The instructional treatment that I have described would likely not have been as effective if students had been unable to reliably determine the number of cubes that fill actual boxes.

REFERENCES

Battista, M. T., & Clements, D. H. (1996). Students' understanding of three-dimensional rectangular arrays of cubes. *Journal for Research in Mathematics Education*, *27*, 258–292.

Ben-Chaim, D., Lappan, G., & Houang, R. T. (1985). Visualizing rectangular solids made of small cubes: Analyzing and effecting students' performance. *Educational Studies in Mathematics*, *16*, 389–409.

Cobb, P., Wood, T., & Yackel, E. (1990). Classrooms as learning environments for teachers and researchers. In R. B. Davis, C. A. Maher, & N. Noddings (Eds.), *Constructivist views on the teaching and learning of mathematics. Journal for Research in Mathematics Education* Monograph Number 4 (pp. 125–146). Reston, VA: National Council of Teachers of Mathematics.

Hirstein, J. J. (1981). The second national assessment in mathematics: Area and volume. *Mathematics Teacher*, *74*, 704–708.

TEACHER APPROPRIATION AND STUDENT LEARNING OF GEOMETRY THROUGH DESIGN

Cathy Jacobson and Richard Lehrer, University of Wisconsin—Madison

Abstract. In 4 Grade 2 classrooms, children learned about transformational geometry and symmetry by designing quilts. All 4 teachers participated in professional development focused on understanding children's thinking in arithmetic. Two of the 4 teachers participated in additional workshops on students' thinking about space and geometry, and they elicited more sustained and elaborate patterns of classroom conversations about transformational geometry than did the other 2 teachers. These differences were mirrored by students' achievement differences that were sustained over time. We conclude that specific subject-matter knowledge, not simply general knowledge about students and their thinking, is needed to support effective teaching.

*P*rinciples and Standards for School Mathematics (National Council of Teachers of Mathematics, 2000) advocated a greater role for geometry and space in the primary grades, a call reflected in recent curriculum development efforts. In this chapter we describe a second-grade curriculum unit in which the students partitioned geometric forms and transformed these forms to produce novel designs for quilts. They predicted outcomes of transformations and compositions of transformations, compared them to the results achieved from physical manipulation, and revised their predictions when necessary. The students developed con-

jectures that were sometimes refuted by themselves or others and invented mathematical theories that accounted for the relationships between conjectures and refutations.

The students first designed *core squares*, squares composed of different shapes (initially right triangles), which served as basic units for quilt designs (see Figure 10.1). After composing core squares, the students employed slides, flips, or

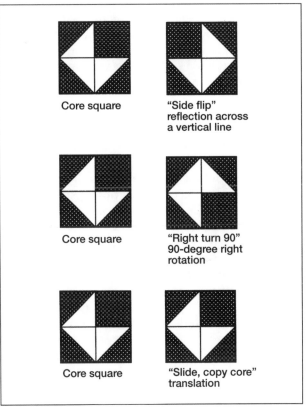

Figure 10.1. Sample core square and three possible transformations.

The first author, Cathy Jacobson, is now deceased.

This chapter is adapted from Jacobson, C., & Lehrer, R. (2000). Teacher appropriation and student learning of geometry through design. *Journal for Research in Mathematics Education, 31,* 71–88.

turns to create a 2×2 design, which in turn was subjected to further transformation or composition of transformations to create a quilt. These design activities were supported by computer software, Geometry in Design (Watt & Shanahan, 1994, pp. 21–64), for practicing rotations, reflections, and translations and for electronic composition of virtual quilts. This design-oriented approach embedded learning about transformational geometry, color composition, and symmetry within meaningful activity.

However, no matter how innovative, curriculum comprises only part of the script of classroom learning: Teachers orchestrate how the curriculum "plays" in the classroom, with attendant consequences for student learning. Although classroom implementation of curriculum is governed by many factors, teacher knowledge of studen' thinking has a consequential effect on instructional decisions (Fennema & Franke, 1992). But some debate arises about how specific a teacher's knowledge of student thinking needs to be to serve as an effective guide. On the one hand, teachers who learn to attend to students' thinking in one area of mathematics (e.g., arithmetic) may adapt that stance to another area of mathematics (e.g., geometry) with little loss of attunement to students. On the other hand, a teacher's knowledge of students' thinking may be more finely structured, so that attunement to students' thinking in one area may fail to result in similar capacity in other areas.

We investigated how four Grade 2 teachers (called A, B, C, and D) taught the unit Geometry in Design. All four were experienced second-grade teachers and had participated in Cognitively Guided Instruction (CGI), a research program in which teachers make instructional decisions on the basis of their knowledge of individual children's thinking (Carpenter & Fennema, 1992). Children in what we refer to as "CGI classrooms" spend most of their time solving arithmetic problems, and teachers use information from each child's reporting of problem solutions to make decisions about how instruction should be structured to foster understanding. We could discern little difference among these classrooms with respect to children's arithmetic problem solving. That is, all four teachers elicited students' strategies for solving a wide range of arithmetic word problems and made instructional decisions on the basis of their knowledge of typical trajectories of change in these strategies.

Two of the teachers (A and B), however, were third-year participants in a research program that offered workshops and other forms of professional development on understanding students' thinking about geometry (Lehrer et al., 1998). In the first year, these teachers (and others) participated in workshops led by a researcher and an elementary school teacher who had extensive experience with CGI. Approximately half (30 hours) of the workshops were devoted to having teachers solve geometry and measurement problems. These workshops focused on how primary-grade students usually solved these types of problems. The teachers saw videotapes of students' solutions. Participating teachers talked about ways to mediate discussions about space, much in the way that they mediated discussions about number. Transformational geometry was not the subject of any workshop.

As a follow-up to the initial workshops, Teachers A and B met with a small cadre of other participating teachers each summer for the 2 years before they participated in this research. Each of the participating teachers brought to the summer project knowledge about how her or his students developed spatial reasoning as a result of the activities they had tried in their classrooms. Jointly, they authored or adapted curriculum tasks that were designed to provide windows to students' reasoning and to "spiral" toward increasingly sophisticated spatial-problem solving. Thus, teachers systematically enlarged their bodies of knowledge about students' thinking by reflecting about what they had learned in practice each year and by revising or authoring new curriculum tasks.

Prior to this investigation, one of our conjectures was that because all four teachers were experienced CGI teachers and were likely to base instructional decisions on students' thinking, differences in students' learning among the four classrooms might be small. On an independent assessment given in a previous year, students' achievement was slightly higher in the classes of the two teachers who were not part of the geometry research program, so, if anticipating any differences, one might expect students in those classrooms to learn a little more than students in the two geometry classrooms. An alternative conjecture was that the two teachers with greater knowledge of student thinking about space and geometry might teach the curriculum differently, with corresponding differences in student learning. To assess these conjectures, we collected two types of data. First, teachers' practices for eliciting and elaborating mathematical talk were observed during periodic classroom conversations about a video depicting quilt design. Second, students' knowledge of transformational geometry was assessed immediately following, and one month after, instruction.

OUR STUDY

Using the Geometry in Design unit, students were engaged in quilt design through hands-on construction of paper quilts, use of interactive computer software, and discussion of a video that depicts quilt designs and computer animation of transformations. The teachers spent 3 to 4 weeks on the unit. Initially students used paper triangles of two colors to compose geometric designs on squares. Then they used rotations, reflections, and slides to create four squares in a 2×2 design. The students used transformations of the 2×2 design to create larger quilt patterns on paper. The students also investigated related mathematical ideas, such as symmetry and quilt "families" (quilt designs that can be made from a single core square). The accompanying video was shown at several points in the curriculum and provided focus for classroom conversation about transformations and various types of quilt design.

Assessing Classroom Discourse

Throughout the unit, the classroom teachers and the students conversed occasionally about what they found interesting or noteworthy in the 5-minute videotape accompanying the unit. The video showed images of quilts, quilt designs, and animation of the transformations used to create the designs but had no spoken narrative. The animations of designs were often accompanied by captions labeling key parts of the design process, for instance, *flip, 2 × 2,* and *core*. The curriculum outline called for the teacher to show the video and to ask students to describe what they saw. Transcribed audiotapes of these discussions were collected and analyzed for types of teacher and student discourse.

Because classroom discussions following the first viewing of the video and a second viewing approximately 2 weeks later occurred at similar points in every teacher's instructional sequence (beginning and midpoint), we selected these tran-

scriptions for further analysis. Here we illustrate only the differences found at the midpoint of instruction. At this point in the unit, the curricular emphasis was on making strip quilts. (Strip quilts are constructed by making the first strip by performing a pattern of transformations on a core square, then flipping or sliding the entire linear strip to make the next row of the quilt.) The teachers were able to use the students' comments about the video as an informal assessment of students' understanding of the geometry of quilt design.

Classes A and B

Teacher A began the discussion by highlighting the quilting terminology that the class had been using to describe what they saw: *turns* and *flips, core squares, 2 × 2s, quilt families, strips*. She continued to focus the students' attention on important mathematical issues by asking them to clarify their statements. This teacher's persistence in identifying students' thought processes and her requests for clarification kept the discussion focused on concepts related to geometric transformations and the implications of transformations for design. In this way, through the discourse, the students elaborated ideas about space instead of merely making a series of observations:

Ned: They're made out of core squares.

Teacher A: How many core squares? Did you happen to notice?

Ned: I think five or six.

Teacher A: Okay, like either five or six, which are made into a strip. How were the strips used to make the quilts?

Ned: They were flipped.

Teacher A: Okay, so sometimes they would—what kind of flip, an up-down flip or a side flip? Do you happen to remember?

[Ned uses his hand to demonstrate an up-down flip as an answer to Teacher A's question.]

At the beginning of the discussion on strip quilts, Teacher B asked students to explain the process of making a strip quilt. She used a student's comment about different designs from one strip to ask the class *why* and *how* questions that prompted hypothesizing about how transformational geometry could be used to change the design.

Jan: They made different designs from the same strip.

Teacher B: Yeah. They made three different kinds of designs from the same strip, didn't they? That was interesting. They started with the same strip. At one time they showed you just copying that strip six times, and it gave you one kind of design, but then they showed you the same strip. They didn't make a new strip, but they showed you the strip and then a flip of the strip and a flip of the strip and a flip of the strip. How was that second design different? What do you think? Nick?

Nick: It kept getting flipped.

Teacher B: And how did that change the design? Right. We flipped the strip.

Student: The other one just kept on going down if you would write on the back just, say, "Copy core, copy core, . . ."

Classes C and D

In the discussion in Class C, the teacher typically responded to a student's comment by asking him to clarify his thinking about turning or flipping the strip to make the "zigzag" quilt. The interaction, however, involved only a student's statement, a teacher's request for clarification about process, and a student's reply. Teacher C rarely followed the initial request for clarification with requests for further elaboration, choosing instead to move on to another student's comment:

Kevin: They used a long one that—it was about . . . ; it was different than the other one. They used a strip that was really long. And they turned it so that there were zigzags.

Teacher C: Did they turn it, or did they flip the strip?

Kevin: They flipped it. And they made zigzags in a longer one.

Teacher C: Um hum. Okay, good idea. Somebody else that wants to share?

In Class D, the teacher prompted the students to discuss what they had noticed on the video about strip quilts. Their discussion, however, focused more on the recognition of shapes and colors than on the geometry of quilt design:

Teacher D: Anything else that you noticed that you hadn't noticed before? Warren?

Warren: From the first and second time I noticed more; I got two more of the core squares.

Teacher D: You could see more of the core squares. How many of you felt that you could see more of the core squares this time because you have worked with them? Okay, a little more than half of you? Okay, good. Any other observations that you want to add?

What can we conclude?

Teachers A and B, experienced at facilitating students' thinking about geometry, sustained longer classroom discussions, both at the beginning of the unit and at the midpoint of the unit. Further analysis of the classroom talk indicates that forms of teacher mediation also differed across the four classrooms. Whereas Teacher A relied primarily on statements intended to focus or clarify students' thinking, Teacher B questioned students to invite conjecture and explanation. Teacher C also invited students' conjectures and helped clarify students' thinking, but she directed the discussions toward the *content* of the core square not toward its mathematical *function*. Teacher D elicited student talk about the video but did not mediate the discussions with either technique.

The statements that involved teachers' revoicing of students' comments represent a relatively small but significant portion of the statements made during the discussions following the quilt video. Each instance of student participation that was fostered and scaffolded by the teacher represented an opportunity for students to learn or reflect upon their learning. As shown in the examples above, teachers' revoicing was differentiated not only by type but also by the depth with which the teacher sustained the discussions about the geometry of designing quilts. in contrast with teachers in the other two classrooms, Teachers A and B used the revoicing techniques to scaffold student talk in ways that invited more precision or reasoning about the causes of what they observed on the video, and that technique or practice extended the discussions about the function of the core square and the motions used to create the quilts.

Assessing Student Knowledge of Transformational Geometry

The students responded to eight problems involving flips, turns, and composition of motions. The first five items measured students' *comprehension* of transformations. Each item required a written response to a transformation demonstrated with color transparencies on an overhead projector. Students wrote about "what would tell someone else in another room the kind of moves you have just seen with the colored squares on the overhead projector." The items in Part 1 (see Figure 10.2) consisted of two rotation items (90° left turn and 135° right turn), two reflection items ("side" flip and an "up-down" flip), and a combination of a rotation and a reflection (90° right turn followed by an up-down flip). Each motion was demonstrated at least twice, with repetitions of the transformation shown to those students who requested it. All students finished their descriptions before the teacher proceeded to the next item.

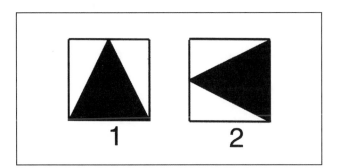

Figure 10.2. Example of an item measuring students' comprehension of transformation.

Each of the five comprehension items in the first part of the assessment was coded (a) for description of the type of motion (e.g., a turn), (b) for direction (e.g., right), and (c) if the motion was a turn, then also for amount (e.g., 90°). The first five items were coded as *wrong* (0 points) or *right* (1 point) for each subcomponent of (a) type, (b) direction, and (c) quantity (if necessary). For exam-

ple, a rotation problem could be scored a maximum of 3 points; a lower score (2 points) was given if the amount of the turn was incorrect. The maximum possible score for the comprehension portion of the assessment was 16 points.

The remaining three items involved pretransformational and posttransformational images. Students needed to specify unambiguously how to transform one core square into another. Each item was scored to indicate the total number of motions or the combination of motions that described a valid transformation. Many students were able to describe a motion in general terms, such as "right turn and then flip down." Students in all four classrooms used the same terminology for describing the motions because they had all used the same curriculum. Some students, however, did not specify all necessary information, for example, the amount of a turn. Unless their descriptions contained the necessary type, direction, and amount (if needed), however, these responses were not coded as valid. Thus, the score for each item in the second part of the assessment consisted of the total number of complete descriptions of how to transform the preimage into the postimage. The number of complete solutions for the three production items ranged from 0 to 5 because some students were able to describe several accurate transformation patterns for a pair of squares.

We tested students' knowledge of transformational geometry at the end of instruction and one month following the conclusion of instruction. The same items were administered both times but were presented in different orders, and different colors for the core squares were used to assess students' understanding of motions.

What the Students Learned

Table 10.1 displays the class mean subscores on both parts of the assessment. The differences between the scores for Classes A and B and those for Classes C and D were evident across the range of motions and were manifested both when motions were modeled (Part 1 of the assessment, measuring comprehension) and when they were not (Part 2 of the assessment, measuring production). The largest differences between the two pairs of classes were associated with composition of transformations and production of transformations, arguably the most challenging items. Furthermore, these differences were maintained over a month-long interval.

DISCUSSION

The teachers of the four classes involved in this study were all skilled and experienced teachers working with a thoughtfully designed sequence of tasks and computer tools to assist students in their learning of transformational geometry. As expected, students in every class comprehended and produced rotations, reflections, and compositions of motions. The teachers of all four classes attended to students' thinking as well as to students' actions during the course of the curriculum. Nevertheless, the trends in the data indicated differences in student achievement. In the two classes (A and B) in which the teachers were more knowledgeable about students' thinking about space and geometry and had therefore appropriated this curriculum in ways that Teachers C and D had not, the students not only learned more than did their counterparts, but this difference in learning was maintained over time. This finding indicates the benefits of teachers' having knowledge attuned to nuances of student thinking within a mathematical domain.

We can attribute these differing profiles of student achievement partly to the teacher orchestration of classroom talk. Although all the teachers elicited students' thinking, the two teachers who were more knowledgeable about students' thinking about space orchestrated classroom talk in ways that refined, elaborated, and extended students' thinking, albeit in different ways. By posing questions and revoicing students' comments that focused, refined, or "lifted out" important ideas, Teacher A orchestrated children's talk about what they saw in the quilt video. During the first viewing,

for example, she highlighted a student's comment ("From just one core square? They can make all those designs from just one core square?") by noting, "Sure, it depends on how you turn it, flip it, rotate it, what you do with it, or the colors that you use."

Teacher B, in contrast, posed questions that promoted students' conjectures about causes of observed patterns. During a discussion following a viewing of the video, she asked the class, "Why does it help you to know that there's lots of different patterns?" This question led to students' explorations of how different transformations affected the same core square and how repetition of simple transformations resulted in "cool" designs. In several instances she appeared to orchestrate a metadiscussion of the intentions of the video producers.

Students in Classes A and B not only learned more about transformations than those in Classes C and D but also retained their knowledge over time. Teacher C used some of the same questioning techniques used by Teachers A and B, although she employed them less often (especially at the midunit point) and dealt with qualitatively different issues (content instead of function). These differences may explain why the students in Class C, who scored lower on the initial assessment, nevertheless demonstrated long-term retention of what they did learn. The students in Class D also learned much about transformational geometry, but their knowledge displayed the decay found in other studies of long-term memory. One reason for the decay may be that Teacher D's style of mediating classroom talk did not include either of the revoicing techniques noted in the other classrooms.

Table 10.1

Class Mean Subscores on Written Assessments of Students' Understanding of Transformational Geometry

			Class		
Assessment		A	B	C	D
End of unit					
Part 1:	Rotation	3.77	3.94	2.67	3.54
	Reflection	3.78	3.62	2.89	3.08
	Composition	4.33	5.25	3.72	3.92
Part 2:	Production	1.56	2.06	0.50	1.00
One month later					
Part 1:	Rotation	3.61	4.25	2.89	2.92
	Reflection	3.56	3.88	2.78	2.15
	Composition	4.56	5.06	3.61	3.00
Part 2:	Production	1.50	2.31	0.39	0.31

Apparently, teacher revoicing of students' talk played an important role in developing students' knowledge. In short, Teachers A and B, who were more knowledgeable than Teachers C and D about research on the ways students reason about space, were better able to use student talk as a means of elaborating students' understanding.

These results underscore the importance of providing opportunities for ongoing professional development that is attuned to specific benchmarks and indicators of students' reasoning about mathematics. Such specificity sets the stage for teachers' construction of models of students' thinking. These models must be of sufficient generativity and flexibility to encompass the ebb and flow of students' reasoning within the dynamic of classroom discourse. Our findings further underscore the fact that mathematics curriculum includes more than tasks and activity; mathematics curriculum includes the interaction between the nature of mathematical discourse and the teacher's understanding of student thinking.

REFERENCES

Carpenter, T. P., & Fennema, E. (1992). Cognitively guided instruction: Building on the knowledge of students and teachers. *International Journal of Educational Research, 17,* 456–470.

Fennema, E., & Franke, M. L. (1992). Teachers' knowledge and its impact. In D. A. Grouws (Ed.), *The handbook of research on mathematics teaching and learning* (pp. 147–164). New York: Macmillan.

Lehrer, R., Jacobson, C., Thoyre, G., Kemeny, V., Strom, D., Horvath, J., Gance, S., & Koehler, M. (1998). Developing understanding of space and geometry in the primary grades. In R. Lehrer & D. Chazan (Eds.), *Designing learning environments for developing understanding of geometry and space* (pp. 169–200). Mahwah, NJ: Erlbaum.

National Council of Teachers of Mathematics. (2000). *Principles and standards for school mathematics.* Reston, VA: Author.

Watt, D., & Shanahan, S. (1994). *Math and more 2—Teacher guide.* Atlanta, GA: International Business Machines Corp.

A LONGITUDINAL STUDY OF INVENTION AND UNDERSTANDING: CHILDREN'S MULTIDIGIT ADDITION AND SUBTRACTION

Victoria R. Jacobs, San Diego State University

Abstract. A 3-year study of development of children's understanding of multidigit addition and subtraction concepts (Grades 1–3) focused specifically on the role that invented strategies play in the development of that understanding. Results indicate that almost all children can and do invent strategies and that this process of invention (especially when it comes *before* learning standard algorithms) may have multiple advantages. Furthermore, use of invented strategies does not preclude use of standard algorithms; the same child often uses both. Conclusions are presented about (a) who could invent strategies and (b) the potential advantages of using invented strategies.

WHEN children solve multidigit addition and subtraction problems, two types of problem-solving strategies are commonly used: invented strategies and standard algorithms (see Figure 11.1). *Invented strategies* are problem-solving procedures that are invented by children rather than shown to them through direct instruction. Invented strategies naturally develop over time as abstractions of children's strategies that are based on tens materials. For example, before inventing strategies, children often begin problem solving by using individual counters to model

This chapter is adapted from Carpenter, T. P., Franke, M. L., Jacobs, V. R., Fennema, E., & Empson, S. B. (1998). A longitudinal study of invention and understanding in children's multidigit addition and subtraction. *Journal for Research in Mathematics Education, 29,* 3–20.

The research reported here was supported in part by NSF Grant Number MDR 8955346. The opinions expressed here do not necessarily reflect the views of the NSF.

addition or subtraction operations. When they gain understanding and experience with larger numbers, they begin to use various materials to represent tens rather than rely solely on individual counters. Finally, children begin to use their knowledge of tens to invent strategies for adding and subtracting larger numbers. These invented strategies are performed mentally without physical materials of any kind.

Standard algorithms, in contrast, are not invented by children but have evolved over centuries for efficient, accurate calculation. Numerals are aligned in columns, and calculations are performed on those columns of single digits. Although this approach simplifies calculations, the procedures can be executed rotely without understanding, and significant ideas can be hidden. For example, some children may not pay attention to which unit (ones, tens, hundreds, etc.) they are adding. The children quoted in Figure 11.1 may not understand that they are adding 2 tens and 3 tens but may instead think that they are adding 2 and 3 simply because they are in the same column. In contrast, most children using invented strategies specifically label the units being combined (e.g., 2 tens and 3 tens, or 20 and 30). A lack of attention to which units are being combined has been linked with various errors, especially in subtraction.

Traditionally, most children have learned standard algorithms. More recently, educators have been exploring children's use of invented strategies. We began this study with the idea that invented strategies might play an important role in helping teachers assess and promote children's under-

	$28 + 36$	$62 - 28$
Standard Algorithms	$\begin{array}{r} 1 \\ 28 \\ +36 \\ \hline 64 \end{array}$	$\begin{array}{r} 5 \\ \overset{1}{\cancel{6}2} \\ -28 \\ \hline 34 \end{array}$
	"Eight plus 6 is 14. Write down a 4 and carry the 1. One plus 2 plus 3 is 6, so 64."	"You can't take 8 from 2, so you have to borrow from the 6. Cross out the 6 and change it to a 5, and change the 2 to a 12; 12 minus 8 is 4, and 5 minus 2 is 3, so 34."
Invented Strategies	1. "Thirty and 20 is 50, and the 8 makes 58. Then 6 more is 64." 2. "Thirty and 20 is 50; 8 and 6 is 14. The 10 from the 14 makes 60, so it's 64."	1. "Sixty take away 20 is 40. Then put back the 2; that's 42. Now take away the 8 from the 42. Take away the 2; that's 40, and then 6 more makes 34." 2. "Sixty take away 20 is 40. You can't take 8 from 2. If you take 2 from the 2, you still have 6 more to take away. Now take the 6 from the 40; that's 34."

Figure 11.1. Sample standard algorithms and invented strategies for addition and subtraction.

standing of multidigit addition and subtraction. Invented strategies are rarely performed rotely without understanding, because children create them on the basis of their *existing* number sense. To investigate this hypothesis, we needed to study children who were in classrooms in which they had opportunities to solve problems in a variety of ways (including using both invented strategies and standard algorithms).

How did we select classrooms to investigate?

We worked with students whose teachers were participating in a 3-year intervention program designed to help them understand and build on children's mathematical thinking. The focus of the program was on how children's intuitive mathematical ideas could assist them in developing understanding of more formal concepts and procedures. Teachers learned about strategies children might use, but they were not given any specific curricular materials or instructional guidelines. Therefore, the teachers' instruction varied. However, several features characterized instruction in most classrooms: extensive use of word problems, freedom for children to solve problems however they wanted (including inventing their own strategies), and dis-

cussions of multiple problem-solving strategies. Although some of the teachers did not formally teach the standard addition and subtraction algorithms, almost all the children had been exposed to the algorithms by the spring of second grade. Thus, all the children in the study came from classrooms that were similar in their focus on word problems and multiple problem-solving strategies (including both invented strategies and standard algorithms). However, the ways that individual children responded to these environments differed. These differences formed the core of our results and helped us understand the potential benefits of using invented strategies. This intervention program has been described by Carpenter, Fennema, and Franke (1996).

How did we select children to follow?

We randomly selected six boys and six girls from each of 10 first-grade classrooms. These classes were drawn from three schools located in predominantly middle-class neighborhoods. We were able to follow 82 of these 120 children through third grade. Over the 3-year period, the children were in classrooms of 27 teachers: 10 in the first grade, 8 in the second grade, and 9 in the third grade.

Lessons Learned From Research

How did we assess each child's strategies and understandings?

We met with each child individually for five problem-solving sessions: winter of Grade 1, fall and spring of Grade 2, and fall and spring of Grade 3. During these sessions, the children were asked to complete problems that assessed—

- their knowledge of fundamental place-value concepts,

- their strategies for solving addition and subtraction word problems and computation exercises,

- their abilities to use *specific* invented strategies, and

- their abilities to use their addition and subtraction problem-solving strategies flexibly.

Figure 11.2 shows examples of problems used. Not all problems were given at each problem-solving session, because some of the easier problems used in the earlier sessions were replaced by more difficult problems in later sessions. Numbers and problem contexts were varied for problems that were used in more than one problem-solving session.

Problem Category	Sample Problems
1. Fundamental knowledge of place-value concepts	• Children were shown a card with *17* written on it and were then asked to count out 17 chips. We pointed to the 7 (in the 17 on the card) and asked children to show with the chips what that part meant. Similarly, we asked them to show what the *1* meant. • Children were asked to solve a word problem that required them to find how many groups of 10 could be made from 36.
2. Strategies for solving addition and subtraction word problems and computation exercises	• Children were asked to solve a variety of straightforward addition and subtraction word problems and computation exercises with two-digit numbers (10–99). In the last two problem-solving sessions, word problems with three-digit numbers (100–999) were also presented.
3. Abilities to use *specific* invented strategies	• We demonstrated how a hypothetical child solved an addition or subtraction problem using an invented strategy. Children were then asked to use the same strategy to solve another problem. *Example:* Children were told that a girl named Sarah solved 65 – 19 by doing 65 – 20 is 45 and then adding 1 back to get 46. Children were then asked to use the same invented strategy to solve 53 – 28. To use this strategy correctly, children needed to round 28 to 30 and then do 53 – 30 to get 23 and then add back 2 (not 1) to get an answer of 25.
4. Abilities to use their addition and subtraction problem-solving strategies flexibly *(Extension problems)*	Children were asked to solve the following problems without counting materials or paper and pencil: • How much of $4.00 would be left over after a purchase of $1.86? • A child has $398. How much more would he have to save to have $500? (missing-addend problem)

Figure 11.2. Sample problems from the problem-solving sessions.

Each problem-solving session lasted approximately 1 hour but was spread over 2 days (30 minutes per day). We read the problems to the children and gave them as much time as they needed to solve the problems. For most problems, the children could select from a variety of materials, including individual counters, base-ten blocks, and paper and pencil. For a few problems, the children were asked not to use counting materials or paper and pencil. These restrictions were included to increase the likelihood that children would use invented strategies if they could do so. For each problem, we asked the child for both the answer and a description of the strategy used to solve the problem.

WHO COULD INVENT STRATEGIES?

Conclusions

- Almost all the children could invent strategies.
- The children used invented strategies for addition more widely than invented strategies for subtraction.
- The use of invented strategies did not preclude a child's use of standard algorithms. The children often used both problem-solving strategies effectively.

We found that most children (88%) used invented strategies at some point during the study. Some children (29%) were already inventing strategies during the Grade 1 problem-solving session. Others did not begin inventing strategies until later in the study. Some children used invented strategies extensively, whereas others used them only occasionally. Note that not just one or two specific problems evoked the use of invented strategies; the children used invented strategies on a variety of problems. Furthermore, use of invented *addition* strategies was more widespread than the use of invented *subtraction* strategies. Specifically, 88% of the children used an invented addition strategy at least once during the study, whereas only 68% used an invented subtraction strategy at least once.

In addition to their widespread use of invented strategies, the children's use of both addition and subtraction standard algorithms was extensive. We found that 99% of the children (all but one child) used the standard addition algorithm at least once during the study. Similarly, 92% of the children used the standard subtraction algorithm at least once. The high percentage of both invented-strategy and standard-algorithm use indicated that these problem-solving strategies can coexist and that children are capable of using both effectively. It also raised the question of whether any advantages are gained by inventing strategies.

WERE ADVANTAGES GAINED BY USING INVENTED STRATEGIES?

Conclusions

We found several advantages for those children who invented strategies before learning the standard algorithms (contrasted with those who learned the standard algorithms first):

- They more quickly gained a deeper conceptual understanding of place value.
- They had a deeper understanding of multidigit addition and subtraction and thus were successful on a wider range of problems (in particular, on extension problems).
- They were more flexible in their abilities to use a variety of problem-solving strategies.
- They made fewer systematic errors.

Before conducting the study, we suspected that using invented strategies before or at the same time as using the standard algorithms might be advantageous. Invented strategies have more obvious links with the underlying mathematical ideas and thus may help children develop deeper understanding of multidigit addition and subtraction.

To investigate the question of whether sequence matters, we clustered the children into groups on the basis of whether they used standard algorithms or invented strategies first. We then looked at how the groups performed on various problems. In particular, we examined group differences in (a) performance on place-value assessments, (b) performance on extension problems, (c) flexibility in strategy use, and (d) incidence of systematic errors during problem solving.

GROUPINGS BASED ON CHILDREN'S INITIAL USE OF INVENTED STRATEGIES OR STANDARD ALGORITHMS

We divided the children into three main groups depending on whether they initially used invented strategies or standard algorithms. (Note that if children used both an invented strategy and a standard algorithm for the first time in the same problem-solving session, they were included in one of the invented-strategy groups. Four children (5%) did not fit into any of these groups. Three of the children never used an invented strategy or standard algorithm successfully in any of the problem-solving sessions, and one child had a unique pattern, using the standard addition algorithm first and an invented subtraction strategy before, or at the same time as, the standard subtraction algorithm.)

- *Invented-strategy group* (33%). The children in this group used invented strategies for both addition and subtraction before, or at the same time as, they used the corresponding standard algorithms.
- *Algorithm group* (22%). The children in this group used standard algorithms for both addition and subtraction before using invented strategies for these operations.
- *Invented-addition-strategy group* (40%). The children in this group used invented strategies for addition before, or at the same time as, they used the standard addition algorithm. However, they used the standard subtraction algorithm before they used invented strategies for subtraction.

Remember that all the children came from classrooms that were similar in their focus on word problems and multiple problem-solving strategies. Therefore, the group differences did not arise because one class was taught the standard algorithm and another class was not. Rather, individual children responded differently to the similar classroom environments, and these differences were of interest to us. Furthermore, although the groups were defined by differences in the order in which the children *first* used invented strategies or standard algorithms, children in these groups showed different patterns of performance throughout the study.

In general, children in the invented-strategy group consistently relied more on invented strategies and less on standard algorithms than did the children in the algorithm group. The invented-addition-strategy group tended to fall between the other two groups in their use of invented strategies and standard algorithms. For example, when examining invented-strategy use in the two Grade 3 problem-solving sessions, we found that 100% of the invented-strategy group, 76% of the invented-addition-strategy group, and only 50% of the algorithm group used an invented strategy at least once during these problem-solving sessions.

We were able to further understand the distinctions among the groups by examining how the availability of paper and pencil influenced each group's use of invented strategies in these two Grade 3 problem-solving sessions. Because standard algorithms (see Table 11.1) require addition and subtraction *in columns*, they can be difficult to perform without paper and pencil. Therefore, some children used standard algorithms regularly when paper and pencil were available and used invented strategies only when they did not have access to paper and pencil. We found that the largest differences were between the invented-strategy and algorithm groups. Only 6% of the algorithm group but 67% of the invented-strategy group used invented strategies when paper and pencil were available.

Examining the influence of the availability of paper and pencil also helped clarify differences between the invented-strategy and invented-

Table 11.1

Percentages of Children Using Invented Strategies in Grade 3 Problem-Solving Sessions

Strategy Use	Strategy group		
	Invented strategy	Invented addition strategy	Algorithm
Used invented strategies *only* when unable to use pencil and paper	33%	61%	44%
Used invented strategies on a variety of problems and even when paper and pencil were available	67%	15%	6%

addition-strategy groups. Although children in both invented-strategy groups tended to use invented strategies relatively early, children in the invented-addition-strategy group more readily abandoned invented strategies for the standard algorithm. For example, by the second problem-solving session (in the fall of Grade 2), 89% of the invented-strategy group and 79% of the invented-addition-strategy group had already used invented strategies for addition. These percentages rose to 100% by the end of the study. By the Grade 3 problem-solving sessions, only 15% of the invented-addition-strategy group but 67% of the invented-strategy group continued to use invented strategies even when paper and pencil were available. Despite their reduced use of invented strategies when given a choice, children in the invented-addition-strategy group were able to use an invented addition strategy when asked specifically to do so. For example, 89% of the invented-addition-strategy group were able to successfully use a hypothetical child's invented strategy for solving a three-digit addition problem (from Problem Category 3 described in Figure 11.2).

Group Differences in Performance on Place-Value Assessments

The children in both invented-strategy groups showed better and earlier understanding of place value than those in the algorithm group. Although we could argue that all multidigit addition and subtraction problems involve some understanding of place value, certain problems (in Problem Category 1 in Figure 11.2) were specifically designed to assess fundamental place-value understanding.

Their performance on these problems indicated that the children in the invented-strategy groups learned fundamental place-value knowledge earlier than those in the algorithm group. By spring of Grade 2, more than 95% of each of the invented-strategy groups had already demonstrated fundamental place-value knowledge. In contrast, only 67% of the algorithm group had done so. These data indicate that invented strategies may provide a context for furthering the development of fundamental place-value knowledge. We also found that having fundamental place-value knowledge was not a prerequisite for children to begin inventing strategies. In fact, 28% of the children in the study used invented strategies *before* demonstrating fundamental place-value knowledge. Thus children may begin inventing strategies early, even when their place-value knowledge is not complete. Furthermore, the *process* of inventing strategies that focus on grouping by tens may assist children in their development of place-value understanding.

Group Differences in Performance on Extension Problems

The children in both invented-strategy groups were better able than other children to use their problem-solving strategies to successfully solve the more difficult extension problems. In both Grade 3 problem-solving sessions, the children were asked to solve two problems (described in Problem Category 4 in Figure 11.2) that required some flexibility in calculating because of the sizes of the numbers, the problem structure, and the restriction that no counting materials or paper and pencil could be used. Children in both invented-strategy groups performed significantly better than children in the algorithm group on the extension problems (see Table 11.2). This difference is important because ability to use knowledge successfully in new situations is an important characteristic of *understanding*. Accordingly, children in the invented-strategy groups showed deeper understanding than did the children in the standard-algorithm

Table 11.2

Average Score for Each Group on the Grade 3 Extension Problems (Maximum Score = 2)

Problem-solving session	Strategy group		
	Invented strategy	Invented addition strategy	Algorithm
Grade 3, fall	1.15	0.88	0.17
Grade 3, spring	1.37	1.15	0.44

Lessons Learned From Research

group, who, in contrast, could not use their knowledge as successfully in this new situation.

Group Differences in Strategy Flexibility

Although children in all groups used both invented strategies and standard algorithms, children in the invented-strategy groups were *more flexible* in their abilities to use a variety of problem-solving strategies. In the algorithm group, only 61% of the children used an invented addition strategy in any of the problem-solving sessions throughout the study, and many of the children in this group could not use invented strategies when specifically asked to do so. For example, on the problem for which they were asked to use a hypothetical child's invented addition strategy, only 39% could do so successfully. These data indicate that many children in the algorithm group were unable to make sense of a problem-solving strategy that differed from their own. In particular, their knowledge base was not comprehensive enough for them to understand using tens and ones in an order that differed from the traditional order of starting with the ones (most invented strategies start with the larger units, such as tens or hundreds). In contrast, the invented-strategy groups were capable of making sense of problem-solving strategies that differed from their preferred methods. Specifically, 70% of the children in the invented-strategy group were able to use the standard addition algorithm by the end of second grade, and almost all of them had used it by the end of third grade. Similarly, by the end of second grade, 91% of the children in the invented-addition-strategy group were able to use the standard addition algorithm.

Group Differences in Systematic Errors

When children with limited understanding of the underlying concepts learn a problem-solving strategy, they often make systematic errors, called "bugs," and they tend to make these errors repeatedly. For example, when using the standard subtraction algorithm, children might subtract the larger number from the smaller number, even when the larger number is on the bottom so that regrouping should be necessary. Children with solid conceptual understanding should not have buggy procedures, whereas children with rote understanding are likely to demonstrate these bugs. In this study, we saw relatively few conceptual errors in children's use of invented strategies, whereas the children exhibited a number of buggy procedures in using standard algorithms. Therefore, we examined the frequency of these systematic errors in the children's use of standard algorithms and found that the invented-strategy group demonstrated more understanding than the algorithm group.

The data showed that some children in all groups demonstrated buggy algorithms and that, overall, subtraction bugs were more common than addition bugs. For example, no child used a buggy algorithm for addition in more than one problem-solving session, but more than 30% of the children did so for subtraction. Significantly more subtraction bugs were found in the algorithm group than in the invented-strategy group. Specifically, 50% of the algorithm group demonstrated subtraction bugs for more than one problem-solving session, whereas only 11% of the invented-strategy group demonstrated these bugs. The invented-addition-strategy group was again between the other two groups, with 38% of the children demonstrating subtraction bugs in more than one problem-solving session. Thus, one could argue that the children using invented strategies gained deeper understanding of multidigit addition and subtraction than their counterparts and that this understanding also helped them to use standard algorithms effectively. This finding is interesting because it underscores the potential value of allowing children to invent their own strategies. Invented strategies not only can be considered a worthy goal themselves but also can support the successful use of standard algorithms.

CONCLUSIONS

This study provides an existence proof, an example that children can invent strategies for adding and subtracting and that this invention may offer several advantages. Compared with children who did not invent strategies, children who did so showed deeper understanding of place value, greater ability to apply knowledge successfully on extension problems, greater flexibility in using a variety of problem-solving strategies (including the standard algorithms), and fewer systematic errors.

Because of the success of children who used invented strategies, one may wonder whether teachers could profitably teach invented strategies directly in much the same way that standard algorithms are currently taught. However, we are skeptical about the direct instruction of invented strategies. If teachers directly taught invented strategies, children might learn them as rote procedures without understanding (as some children currently do with standard algorithms). We suspect that invented strategies may be beneficial partially because they are *invented by children* and are built on their existing understandings.

We began the study with the idea that invented strategies might play an important role in helping teachers assess and promote children's understanding of multidigit addition and subtraction. The data from this study support our assertion and underscore the potential benefits of having children invent their own strategies. The data also show that invented-strategy and standard-algorithm use are not mutually exclusive. Rather, the use of invented strategies can help children develop understanding of multidigit addition and subtraction that can enhance their performance even when algorithms are taught.

REFERENCE

Carpenter, T. P., Fennema, E., & Franke, M. L. (1996). Cognitively Guided Instruction: A knowledge base for reform in primary mathematics instruction. *The Elementary School Journal*, 97, 3–20.

CHAPTER 12

INTERFERENCE OF INSTRUMENTAL INSTRUCTION IN SUBSEQUENT RELATIONAL LEARNING

Dolores D. Pesek, Southeastern Louisiana University
David Kirshner, Louisiana State University

Abstract. To balance their professional obligation to teach for understanding against administrators' push for higher standardized test scores, mathematics teachers sometimes adopt a 2-track strategy: teach part of the time for meaning (relational learning)[1] and part of the time for recall and procedural-skill development (instrumental learning). A possible negative effect of this dual approach is noted when relational learning is preceded by instrumental learning. A group of students who received only relational instruction outperformed a group of students who received instrumental instruction prior to relational instruction.

D EBATES on how to characterize learning and understanding of school mathematics reflect different and perhaps irreconcilable interests of various constituencies within the education enterprise. Those in the administrative infrastructure, reflecting the responsibilities of political and public oversight of education, demand unambiguous documentation of learning. This demand creates pressure to regard knowledge as comprising discrete elements that can be individually assessed and evaluated. In contrast, according to some theories of learning, "the degree of understanding is determined by the number and strength of the con-

nections [among representations]" (Hiebert & Carpenter, 1992, p. 67) rather than by knowledge of unconnected elements.

The brunt of this conflict is borne by the classroom teacher. On the one hand, the organizations in which teachers' professional identities are vested regularly call for teachers to provide instruction that leads to meaningful learning (e.g., America 2000, 1991; National Council of Teachers of Mathematics, 1989, 2000). Instruction should involve students in reflecting, explaining, reasoning, connecting, and communicating. Students should develop relational understanding—"understand[ing] both what to do and why" (Skemp, 1987, p. 9). Mathematics teachers consistently receive this message in journals, at conferences, and during in-service programs.

On the other hand, administrators, parents, and political agents often press for instruction that is based on the belief that students must learn skills first and foremost. Curriculum guidelines list individual topics and skills. Structured time lines and specific textbooks are recommended or required to guarantee that all curriculum topics are "covered." Teachers are encouraged to drill students and have them practice throughout the academic year—particularly during the weeks immediately preceding standardized assessments. Such instruction leads to what Skemp referred to as *instrumental learning,* "learning rules without reasons" (1987, p. 9).

Eisenhart et al. (1993) found that classroom teachers trying to address all these needs are under great pressure. Caught between professional and administrative demands, they frequently adopt a two-track strategy: They spend some time on drill

This chapter is adapted from Pesek, D. D., & Kirshner, D. (2000). Interference of instrumental instruction in the subsequent relational learning. *Journal for Research in Mathematics Education, 31,* 524–540.

[1] The terms *relational* and *instrumental* were coined by the late Richard R. Skemp, to whom we dedicate this report for his contributions to the topic and his encouragement of this study.

and practice to provide for skills and facts and some time on integrating understandings. Because relational instruction is usually assumed to take more time than instrumental instruction to implement, time constraints are often cited as a principal reason for teachers' relying mostly on instrumental instruction.

Because alternating instructional modes is an obvious resolution of the dilemma, exploring effects of this practice on students' learning is worthwhile. Indeed, several mathematics education researchers have reported finding interference effects when initial instrumental learning is followed by relational learning (e.g., Mack, 1990; Wearne & Hiebert, 1988). These researchers suggest that instrumental instruction has a negative influence on subsequent efforts to teach mathematics for meaning. But that finding was only an incidental result of the studies and deserves more careful investigation.

Our study is an extension of the research into interference effects of instrumental learning. We set out to determine whether relational instruction alone could be more effective than relational instruction that is preceded by instrumental instruction. Note that for the purposes of experimental control, we necessarily created exaggerated and narrowly focused learning environments (instrumental or relational) instead of replicating more typical classroom learning experiences. We then reinterpreted our results as they relate to more normal classroom environments.

Interference of Prior Learning

Two sorts of interference of prior learning on subsequent learning have been identified. The more important of these, *cognitive interference*, occurs when previous understandings in a domain are so powerful as to spontaneously affect subsequent learning. The second type of interference we call *attitudinal interference*. In this type the student's previously acquired opinions and attitudes serve to block full engagement in a situation that might otherwise be productive for learning. For instance, a student's low self-efficacy may decrease that student's willingness to engage fully in instructional activities. Or school-based views of mathematics problems as never requiring more than 5 minutes to solve may mitigate against students' persisting with more substantial mathematical problems (Schoenfeld, 1994).

A third type of interference, *metacognitive interference*, in some sense is intermediate between the cognitive and attitudinal varieties. As with cognitive interference, previous content affects learning new content. But unlike in cognitive interference, the previous competence is not firmly entrenched. Rather, for metacognitive interference, the student's initial competence is effortful, not automatic. Indeed, maintaining this competence requires substantial rehearsal or other mental effort, leaving little opportunity for learning the content approached in a different way. That is, instruction in new methods in this content domain can threaten the student's existing competence by drawing away the mental resources needed to maintain it. Therefore, the student basically ignores new instruction. Note that these forms of interference are not mutually exclusive. For instance, negative attitudes toward relational mathematics instruction may reflect a student's discomfort with the excessive cognitive demands required to simultaneously maintain instrumental competencies.

METHOD USED FOR THIS STUDY

Our objective in this study was to compare the effects of students' receiving instrumental instruction prior to relational instruction (the I-R treatment) with the effects of their receiving relational instruction only (the R-O treatment). In designing the treatments, we decided to provide a longer period of time (5 days) for the instrumental instruction than for the relational instruction (3 days). This decision was based on two considerations. First, we believed that this ratio reflects the situation in many classrooms in which relational methods are employed as "extra" activities within a classroom culture generally dominated by instrumental instruction. Second, using a shorter relational treatment provided the possibility of highlighting the effectiveness of relational teaching, challenging the usual assumption that meaningful instruction requires more time for addressing the same content.

Students from 6 fifth-grade mathematics classes participated in this study; three classes were taught by each of 2 fifth-grade mathematics teachers in a middle-class semirural school. All these classes were grouped heterogeneously for mathematics instruction. We separated each class into two treatment groups using random stratification by gender and achievement level. Achievement level was measured by California Achievement Test (CAT) scores.

The mathematical content chosen for this study was area and perimeter of squares, rectangles, triangles, and parallelograms. This content was selected because it was believed to be conceptually accessible to the students and because formulas for calculations had not yet been introduced to the students. (Instruction in formulas for area and perimeter was scheduled for the end of the fifth grade.)

All the students first received a written pretest. Half of the students (the I-R group) then received 5 days of instrumental instruction. Both groups of students, the I-R group and the R-O group, then received 3 days of relational instruction on the same content. Both groups received a posttest following the relational instruction and a retention test 2 weeks later.

The Two Instructional Treatments

The formulas taught instrumentally included those for perimeter and area of squares ($P = 4s$, $A = s^2$), rectangles ($P = 2[l + w]$, $A = LW$), triangles ($P = a + b + c$, $A = 1/2bh$), and parallelograms ($P = 2[b + w]$, $A = bh$). The instrumental instruction was carefully designed to facilitate the I-R students' memorization and routine application of the formulas. A typical day's lesson for the instrumental instruction started with a review of formulas previously learned. Then a shape was drawn; the dimensions were labeled with appropriate variables; and the formula for finding the desired measure was written on an overhead transparency. The students were asked to write the new formula 10 times. The instructor then explained the indicated operations in the formula and connected the variables in the formula with the corresponding features of the shape. Values for the variables were assigned, and the teacher demonstrated the use of the formula for finding the desired measure. The students worked three problems with the instructor and then five additional problems in cooperative groups. The period ended with a quick review of the formulas presented that day. At no time were the formulas justified in terms of the characteristics of the geometric figures to which they applied. During these 5 days, the R-O group remained with their regular teacher and reviewed material unrelated to the content of this study.

In the relational instruction, we tried to use students' intuitions about size and distance to develop measurement strategies for finding area and perimeter in the context of these geometric shapes. The instruction consisted of three 1-hour lessons given to intact classes (consisting of both the I-R and R-O students) over a 3-day period. We designed the instruction to encourage students to construct relationships. Area and perimeter for each shape were presented together to help students contrast and compare these constructs. The shapes were discussed in the following order: squares, rectangles, parallelograms, and triangles. Connections were developed through concrete materials, questioning, student communication, and problem solving. We designed this instruction to assist students in constructing their own ways to calculate measures of area and perimeter on the basis of their understanding of these two concepts. The teacher initially presented each problem to the whole class to ensure that all the students understood what was required. Then the students worked in cooperative groups of four to solve the problem. Finally the class as a whole discussed the solutions and strategies used by the groups.

The teacher of the relational instruction never specified strategies for efficient or effective calculation. Initially most students used counting strategies to find the area or perimeter of the given figure. Because small numbers were used in the initial problems, such counting strategies could be used effectively. But when the sizes of numbers given in the problems increased, students had greater difficulty counting the area and perimeter units. We used the larger numbers to encourage the students to develop more sophisticated methods for calculation. Gradually most students used increasingly sophisticated methods, although a few did not progress beyond using counting strategies. In all cases, the strategies students used evolved out of their intuitive understandings of area and perimeter and were sensible to them. No pencil-and-paper calculations were taught during this relational unit.

In addition to carefully attending to these pedagogical practices, we designed the instructional sequence following a careful analysis of the conceptual difficulties posed by specific geometric figures. For instance, triangles were addressed only during part of the last day because area calculations for triangles are complex.

RESULTS

Comparing the Groups' Performance on Tests

The pretest, posttest, and retention test given to all the students were nearly identical. These instruments consisted of 34 open-ended items and incorporated a variety of kinds of test items to evaluate a student's ability to calculate the areas and perimeters of shapes. Twenty-four of the items on the tests could be solved using one of the formulas presented in the instrumental instructional unit. The other items could *not* be solved by straightforward use of the formulas presented in the instructional unit: Seven of these items gave the perimeters or areas (and, when needed, one dimension) and required the student to derive a dimension (e.g., find the measure of each side of a square, given the perimeter). The remaining items involved complex shapes constructed by adjoining two or more simple figures (triangles, squares, and rectangles), for example, an L-shaped figure formed by adjoining two rectangles. Four of these items involved determination of the areas of the complex figures. Two of these items involved determination of the perimeters of the complex figures.

In scoring the pretest, posttest, and retention test, we assigned each correct response a value of 1 point. The maximum score was 34. The criterion for scoring the responses was whether the student displayed an understanding of area or perimeter, as required. We used statistical measures to check the reliability of each test. We then used a statistical test to look at differences between groups on the posttest and retention test. We found that the pretest and the CAT both significantly influenced the posttest and retention test.

For students receiving only relational instruction, the posttest mean score was 16.42 and the retention-test mean score was 16.49, in comparison with a posttest mean score of 14.36 and a retention-test score of 14.31 for students who were given instrumental instruction prior to relational instruction. Statistical tests were used to determine whether these differences in group performance were a result of the treatment differences or just random chance variations. These tests indicated about a 94% likelihood that treatment differences caused the performance differences—less than the 95% likelihood generally required to validate a result. However, our sample size was less than optimal. A subsequent test of the power of the design showed that if the same pattern of results had been achieved with a larger sample, the 95% criterion would have been met. We also found that stronger students seemed able to overcome the negative effects of instrumental instruction more easily than weaker students.

Comparing Groups Through Interview Analyses

In addition to using quantitative information from the tests, we interviewed six students from each group. In each group of six interviewees were three boys and three girls; three of the students were high achieving, and three were low achieving. The interview data supplemented the quantitative data to provide us with a better understanding of any interference effects that might be revealed in the test scores. Each student was interviewed three times; here we discuss only the final-interview results. The interview questions from the final interview addressed the students' feelings about the two types of instruction, their understandings of area and perimeter, their abilities to apply knowledge of area and perimeter, how students solved particular problems on the posttest, and how students interpreted formulas for perimeter and area. We identified cognitive, attitudinal, and metacognitive characteristics of the interference of instrumental learning.

Cognitive interference

The difference in conceptual grounding of the students was apparent in their responses to some of the tasks posed in the interviews and seems to partly explain the differential success rates of the two groups. For example, when asked a question regarding the amount of paint or wallpaper needed for the walls, five of the six I-R interviewees said either that they did not know or that one needs to know the perimeter of the room to determine the amount of paint or wallpaper required for the walls. A typical explanation was "Walls don't have area because they go around." In contrast, all six of the R-O interviewees recognized that area measurements are needed to determine the wall's surface.

Differences between the students in the two groups also emerged in their understandings of dimensions and dimensionality in discussion of real-world applications of area and perimeter. The I-R students mistakenly claimed that area is needed to measure "the lengths of boards," "how much liquid," "the thickness of concrete," and the "height of a pole." R-O students made no such errors in their applications of area to life situations. The concepts of area and perimeter appear to have been less clearly differentiated for the I-R students than for the R-O students.

These differences contributed to students' varying appreciation of the possible applications of this mathematical content to their out-of-school lives. In discussing who needs to understand area and perimeter and why they need to understand those concepts, the R-O interviewees gave a greater number of concrete applications (for carpet, painting, wallpaper) than the I-R students did (for tests, later study, college).

Attitudinal interference

The experience of relational instruction was new for most of the students. They all agreed that their regular mathematics instruction was very different from the instruction in this study. Perhaps because of the novelty, all those interviewed said that they enjoyed the relational unit, but students in the I-R group varied in their preferences for the two forms of instruction. Half of them stated that the formula instruction was easier and more enjoyable, and all but one thought that they had learned more from the formula instruction. The one student preferring the second treatment complained, "They [the formulas] got me confused. It's complicated to remember all the stuff about which formula goes with which problem." Perhaps the similarity of the instrumental instruction to their regular classroom practices influenced students' greater receptivity to the instrumental component.

Metacognitive interference

The interviewees' explanations of how to calculate area and perimeter reflected very different approaches to the subject matter. The I-R students consistently used operations (usually incorrectly or with incorrect reasons) in their approaches to calculating area and perimeter. For instance, when asked how one can find area, an I-R student said, "Sometimes you multiply; sometimes you add." For the area of a 2-by-2 shape, the student recommended multiplication, "2 times 2"; but the student gave a relationally incoherent reason: "Because there's four sides." One I-R student demonstrated understanding at the start of her explanation of perimeter but seemed to then apply operations at random by stating, "It's like when you have a rectangle or some-

thing, and if you have four on one side and three on one side, …you walk around it and you count how [much] distance… And then sometimes they will multiply it and get the answer, or they will add it and then they will add the two and then they will multiply the other ones with it." The R-O interviewees' general approaches tended to be grounded in the concrete methods used in the relational instruction ("use hands," "use books," "use tiles" to cover the surface). The solution methods they used during the interviews involved quantitative reasoning about diagrams constructed or provided during the interviews.

The I-R interviewees' ungrounded use of formulas carried over into their explanations of how formulas work. When asked to explain why the perimeter of a rectangle and the area of a triangle are given by the formulas $P = 2(l + w)$ and $A = 1/2\ bh$, respectively, all I-R students either said that they did not know or gave incorrect explanations. For instance, no one in this group could correctly explain the role of the 2 in the perimeter formula. Some erroneous explanations were "the length is 2 and the width is 2," "it has 2 numbers," "because it has 2 different sides," and "it makes it easier to remember." In explaining the triangle-area formula, one of the I-R students simply said that he did not understand, and four students gave confusing explanations (e.g., "I think one measure is one half the other"; "you take the number and you add them together"; "you take one half away"; and "one half [is there] because the sides are not always going to be the same"). One displayed the alchemist's initiative in explaining "there's three sides on a triangle [therefore] … there's three numbers."

Interestingly, the R-O interviewees were able to make more sense of the formulas, despite their lack of instructional exposure to them. In some cases R-O students reported that they simply could not see any connections. However, five of the six students formulated partial or complete explanations for formulas for at least one of the figures, and no explanation given was incorrect. "For $P = 2(l + w)$, one needs to add the length and width and then double it" was the gist of statements by several R-O students. In various ways they explained that the sides were repeated twice. The students who were able to connect the area of a triangle with its formula explained that a triangle is half of a rectangle.

The fact that I-R interviewees relied heavily on formulas but understood them poorly reveals conditions under which metacognitive interference might be expected to arise. Maintaining applicative skills through memorization requires concerted mental rehearsal and other attentional resources. To the extent that students are committed to maintaining these applicative skills, engaging with subsequent relational instruction might be experienced as a distraction. Engaging less fully with the relational instruction would lessen this distraction.

DISCUSSION

This study, like earlier studies (e.g., Mack, 1990; Wearne & Hiebert, 1988), indicates a danger in the compromise of teaching for rote-skill development part of the time and for conceptual understanding part of the time: Initial rote learning of a concept can create interference with later meaningful learning. In this study, students who were exposed to instrumental instruction prior to relational instruction achieved no more, and most probably achieved less, conceptual understanding than students exposed only to the relational unit. Contrary to the common-sense expectation regarding time-on-task, more instruction does not necessarily translate into more learning.

Students in the two treatment groups emerged with very different approaches to solving problems involving area and perimeter. The interviewees who had received both instructional treatments referred to formulas, operations, and fixed procedures as the means for solving problems. Students who had received only the relational instruction used conceptual and flexible methods of constructing solutions from the units of measurement with which they had had concrete experiences.

The rote-instruction unit was almost twice as long as the relational-instruction unit. This situation replicates the circumstance that we believe characterizes many mathematics classes: Most instructional time is spent on routine exercises to consolidate rote or procedural knowledge; much less emphasis is given to students' intuitive and sense-making capabilities. Thus the interference effects suggested here in the experimental microcosm may reflect the learning experiences of many students in many classrooms.

These results may be most relevant for those teachers who do attempt to respond to the recommendations of the professional organizations by incorporating relational learning activities but maintain a classroom culture dominated by instrumental learning. The hard truth is that real reform in mathematics education calls for a reorientation of classroom norms and practices (Cobb, Boufi, McClain, & Whitenack, 1997; Lampert, 1990). Relational units appended to the existing classroom regimen may effectively be blocked from achieving the relational understanding sought. We hope that the relative efficiency and effectiveness of relational instruction that is not preceded by instrumental instruction, demonstrated in this study, inspires other teachers to undertake fundamental classroom reform.

REFERENCES

America 2000. (1991). *An educational strategy to move the American educational system ahead to meet the needs of the 21st century*. Washington, DC: U.S. Department of Education.

Cobb, P., Boufi, A., McClain, K., & Whitenack, J. (1997). Reflective discourse and collective reflection. *Journal for Research in Mathematics Education, 28*, 258–277.

Eisenhart, M., Borko, H., Underhill, R., Brown, C., Jones, D., & Agard, P. (1993). Conceptual knowledge falls through the cracks: Complexities of learning to teach mathematics for understanding. *Journal for Research in Mathematics Education, 24*, 8–40.

Hiebert, J., & Carpenter, T. P. (1992). Learning and teaching with understanding. In D. A. Grouws (Ed.), *Handbook of research on mathematics teaching and learning* (pp. 65–97). New York: Macmillan.

Lampert, M. (1990). When the problem is not the question and the solution is not the answer: Mathematical knowing and teaching. *American Educational Research Journal, 27*, 29–63.

Mack, N. K. (1990). Learning fractions with understanding: Building on informal knowledge. *Journal for Research in Mathematics Education, 21*, 16–32.

National Council of Teachers of Mathematics. (1989). *Curriculum and evaluation standards for school mathematics*. Reston, VA: Author.

National Council of Teachers of Mathematics. (2000). *Principles and standards for school mathematics*. Reston, VA: Author.

Schoenfeld, A. H. (1994). Reflections on doing and teaching mathematics. In A. H. Schoenfeld (Ed.), *Mathematical thinking and problem solving* (pp. 53–70). Hillsdale, NJ: Erlbaum.

Skemp, R. R. (1987). *The psychology of learning mathematics*. Hillsdale, NJ: Erlbaum.

Wearne, D., & Hiebert, J. (1988). A cognitive approach to meaningful mathematics instruction: Testing a local theory using decimal numbers. *Journal for Research in Mathematics Education, 19*, 371–384.

CHAPTER 13

AN EXPLORATION OF ASPECTS OF LANGUAGE PROFICIENCY AND ALGEBRA LEARNING

Mollie MacGregor and Elizabeth Price, University of Melbourne, Australia

Abstract. The association of 3 cognitive components of language proficiency—metalinguistic awareness of symbol, syntax, and ambiguity—with students' success in learning the notation of algebra was investigated. Pencil-and-paper tests were given to assess students' metalinguistic awareness and ability to use algebraic notation. In a total sample of more than 1500 students, aged 11 to 15, who were in their 1st to 4th years of algebra learning, very few students with low metalinguistic awareness scores achieved high algebra scores. Implications of this finding for the school algebra curriculum are discussed.

THE question of whether language proficiency is related to learning ability and general academic achievement has been debated for many years. Much of the debate has centered on the performance of immigrant or ethnic-minority students with limited English proficiency. Confounding factors in empirical studies of language proficiency and academic achievement include bilingualism, ethnicity, socioeconomic status, use of nonstandard dialects, and other social and cultural variables. As Tate (1997) pointed out in his review of quantitative research studies of mathematics performance by various social groups, very little research has focused on the association between language proficiency itself and trends in performance. This research may be lacking because language proficiency is difficult to disentangle from social and cultural factors. The lack may also be due to political considerations. The question of whether language proficiency affects mathematics learning "is a political question as well as an educational question" (Tate & D'Ambrosio, 1997, p. 650). Nevertheless, in his overview of the research on race, ethnicity, social class, language, and achievement in mathematics, research that had been carried out in the United States up to 1990, Secada (1992) found sufficient evidence to conclude that "language proficiency, however it is measured..., is related to mathematics achievement" (p. 639).

In the field of literacy and learning, studies of the language skills of bilingual students led Cummins (1984) to suggest that a certain level of linguistic proficiency seems to necessary for academic achievement. This proficiency enables one to use language as an organizer of knowledge and a tool for reasoning. Students learning a second language without a foundation of first-language competence may not develop sufficient proficiency in either language to support their learning. Mestre (1988) referred to language-minority students in this situation as "semilinguals"—casualties of inadequate bilingual programs—and suggested that semilingualism is a cause of their low achievement in mathematics. However, monolingual students too may exhibit low levels of language proficiency and may perform badly in mathematics, and therefore low achievement cannot be attributed to bilingualism as a primary cause.

Much of the research literature on language and mathematics learning is concerned with students' understanding of mathematical information expressed in natural language. In this chapter, how-

This chapter is adapted from MacGregor, M., & Price, E. (1999). An exploration of aspects of language proficiency and algebra learning. *Journal for Research in Mathematics Education, 30,* 449–467.

This research was supported in part by a grant from the Australian Research Council.

ever, we are concerned with a deeper level of language competence—the cognitive level at which symbol processing takes place. We describe our attempt—without precedent as far as we know—to identify components of the ability to process language that may affect the learning of algebra.

Metalinguistic Awareness

In the literature on children's acquisition of literacy, we encountered the term *metalinguistic awareness*. This term has been used by several people in the field of literacy development to refer to the linguistic ability that enables a language user to reflect on and analyze spoken or written language. Metalinguistic awareness is involved when the form or function of a word or phrase, not only its meaning, is the object of attention. An example is a child's noticing that *school* rhymes with *pool* and *rule* and wondering why *school* is not spelled as *skool* and *rule* is not spelled as *rool*. The child is paying attention to the sounds and spellings of the linguistic signs instead of to their meanings.

Metalinguistic awareness enables the language user to reflect on the structural and functional features of text as an object, to make choices about how to communicate information, and to manipulate perceived units of language. Because analyzing structure, making choices about representation, and manipulating expressions are intrinsic to mathematics, and particularly to algebra, it seems likely that metalinguistic awareness in ordinary language has an equivalent in algebraic language. We suggest that this ability in algebra operates at the same level of abstraction as metalinguistic awareness in ordinary language when words and word strings are treated as instances of variables with general properties (e.g., the word *simplify* could be considered as an instance of the variables *string of eight letters, transitive verb, word of three syllables*, or *word starting with s*). An investigation of students' metalinguistic awareness might provide information about specific processes or cognitive structures common to both language proficiency and algebra learning, and it might help explain the difficulties experienced by some students in learning mathematics in general and algebraic notation in particular.

Herriman (1991) listed seven components of metalinguistic awareness in ordinary language. We recognized two of the components, *word awareness* and *syntax awareness*, as having algebraic analogs.

We have translated these two components into the context of algebra.

1. *Symbol awareness* (analogous to word awareness in Herriman's list) includes knowing that numerals, letters, and other mathematical signs can be treated as symbols detached from real-world referents. It follows that symbols can be manipulated to rearrange or simplify an algebraic expression, regardless of their original referents. Another aspect of symbol awareness is knowing that groups of symbols can be used as basic meaning-units. For example, $(x + 2)$ can be considered as a single quantity for the purposes of algebraic manipulation.

2. *Syntax awareness* includes recognition of well-formedness in algebraic expressions (e.g., knowing that $2x = 10$ implies $x = 5$ is well formed, whereas $2x = 10 = 5$ is not well formed) and ability to make judgments about how syntactic structure controls both meaning and making of inferences (e.g., knowing that if $a - b = x$ is a true statement, then it is not generally true that $b - a = x$).

Our working hypothesis was that one or both of these components of metalinguistic awareness are necessary for success in learning to use algebraic notation.

Development of Metalinguistic Awareness

Children who are learning to read exhibit large individual differences in metalinguistic awareness. Researchers have not been able to determine whether metalinguistic awareness is a precursor of learning to read or whether it develops as a consequence of reading. Many researchers have studied the development of children's awareness of structural features of language and their abilities to manipulate language forms. *Symbol awareness*, that is, the awareness that words are arbitrary names and can be represented as groups of symbols, develops during the early primary grades and is associated with learning to read. Symbol awareness is the basis for word games, jokes, and puzzles that are understood and enjoyed by most children. We see no reason that children who are beginning algebra at age 11 or 12 should lack symbol awareness.

In contrast, the development of syntax awareness extends over a longer period and may depend on the development of formal operational thinking (Herriman, 1991). Much everyday reading requires giving very little attention to syntax, relying on context instead. Using contextual clues is not helpful in interpreting algebraic notation, which requires that careful attention be given to the order and arrangement of the symbols. Research has shown that a significant proportion of students beginning secondary school do not distinguish such an expression as 6 – 10 from its inverse 10 – 6 (e.g., Brown, 1981). They use lexical information ("there's a 6 and a 10 and a minus") and then rely on the context (which might be a fall in temperature from 6 degrees or the use of a $10 note for shopping) to choose the answer (in this example, whether the answer should be –4 or 4). They do not reflect on the structure of the symbolic expression itself. Such students have not learned how to use syntax as a guide to interpretation in arithmetic, and they are not likely to understand the significance of symbol order in algebraic notation.

Initial Difficulties in Learning Algebra

In the field of algebra learning, one of the initial obstacles to progress is that students do not easily learn how to express simple operations and relationships in algebraic notation. They frequently misuse and misinterpret algebraic symbols and algebraic syntax, even in such simple tasks as "David is 10 cm taller than Con. Con is b cm tall. What can you write for David's height?" For example, in that task, the belief that algebraic letters are abbreviated words leads to the answer Db (intended to mean "David's height"), and misunderstanding algebraic syntax leads to the answer $b10$ (intended to mean "add 10 to b"). We hypothesized that students' poor understanding of symbols and syntax in algebra may reflect inadequate symbol and syntax awareness in ordinary language.

OUR FIRST STUDY

We explored this hypothesis in a study in which we attempted to test whether an association exists between students' algebra learning and their awareness of symbols and syntax, the two components of metalinguistic awareness listed above. We devised items that tested symbol awareness (e.g., see Item 2 in the Appendix) and syntax awareness (e.g., see Item 3 in the Appendix) and also typical algebra items (e.g., see Item 4 in the Appendix). The test was given to 1236 students of ages 11–16.

We found that the students who obtained maximum or near maximum scores on algebra items also obtained maximum or near maximum scores on language items. Considerable numbers of students with high language scores had low algebra scores. In contrast, no instance occurred, at any of the four grade levels included in the study, of a student with a low language score and a high algebra score. We saw that the language items were not sufficiently difficult to produce a useful spread of scores and concluded that more difficult language items were required in further testing of the relationship between language skills and algebra learning. We therefore decided to undertake another study with a more difficult test measuring metalinguistic awareness and to use it to determine whether metalinguistic awareness is associated with students' success in using algebraic notation.

OUR SECOND STUDY

Because of our experience as mathematics teachers, we propose a third component of metalinguistic awareness in addition to symbol awareness and syntactic awareness; the component does not appear in Herriman's list, and it has a mathematical analog. This component is awareness of potential ambiguity, which, in algebra, is the recognition that an expression may have more than one interpretation, depending on how structural relationships or referential terms are interpreted (e.g., knowing when brackets are required for ordering operations and being aware of the potential for mistranslating relational statements to equations). In the literature on language development are several studies of children's growing awareness of ambiguity when they reach adolescence, an awareness that seems necessary for algebra learning. We conjectured that the use of ambiguity items in our tests combined with more difficult items testing word awareness, such as is called for in understanding metaphors, should result in a more useful spread of language scores. We expected that understanding of metaphor would be an indicator of word awareness because word awareness involves knowing that words are symbols that can be separated from their referents.

Preparation of test and selection of students

We prepared a test containing some new language items provided by teachers of English and some previously used language and algebra items. (The Appendix contains this test.) The new language items had no obvious association with any mathematical concepts, whereas the language items used in our first study required understanding of counting and measuring. These new items were therefore expected to be better indicators of metalinguistic awareness. Examples include items that tested for understanding of metaphor (e.g., see Item 18 in the Appendix) and items that tested for awareness of ambiguity (e.g., see Item 5). This test was given to all students in Grades 8 through 10 in one school in a middle-class suburb of Melbourne. Responses from students whose first language was not English were excluded from the analysis of data. Responses to algebra items were scored either right or wrong (i.e., scored 1 or 0). Responses to language items were scored 0, 1, or 2: 1 point was given for evidence of partial understanding, and 2 points were given for evidence of good understanding.

Results

The correlations between language and algebra scores were found to be positive for each grade level. The use of more difficult language items produced a more acceptable spread of scores. In the new data, no students who had very low algebra scores attained very high language scores. High language scores tended to be associated with high algebra scores. We found few instances of a student's having a very low language score and a high algebra score. This finding runs counter to the popular belief that low ability in language is not necessarily a barrier to high achievement in mathematics.

DISCUSSION

We do not know why, in both Study 1 and Study 2, some students with good language scores got many algebra items incorrect although they had been given the same opportunity to learn as their classmates. One explanation is that these students were not sufficiently aware that the algebraic sign system has its own grammatical rules and conventions that are not intuitively obvious and must be learned. Possibly some students are misled by their experiences with other symbol systems. We find no reason that they should not succeed in learning algebra if they are helped to recognize that changes in the order, position, or grouping of symbols affect meaning and that the language of algebra has its own set of grammatical rules that are not intuitive but must be learned and practiced. For students with low metalinguistic awareness, however, supporting their language development may be a better approach. An important topic for research is whether the development of meta-

linguistic awareness can be accelerated and whether algebra learning should be postponed until an adequate level of metalinguistic awareness has been reached.

Teachers are concerned that some students learn very little of the algebra taught in introductory courses. The evidence presented in this chapter indicates that, for some of these students at least, their failure to learn is associated with poorly developed metalinguistic awareness. This conscious awareness of language structures and the ability to manipulate those structures may be manifestations of deeper cognitive processes that also underlie the understanding of algebraic notation. The existence of a common cognitive basis could account for the association between algebra learning and metalinguistic awareness, indicated by our study. The relationship of metalinguistic awareness with what is commonly called general intelligence or cognitive ability needs to be investigated.

Within the mathematics education community, much discussion centers on the inclusion of algebra in a common curriculum. What algebraic knowledge should be taught in an "algebra for all" course? What is the minimum standard that all students can be expected to achieve? What algebra skills do people need for the 21st century? A report (Stacey & MacGregor, 1997) on algebra learning in the State of Victoria, Australia, where algebra is part of the curriculum for all students in the first 4 years of secondary school, indicated that a certain proportion of students do not develop basic competence in using algebraic notation during those 4 years. If understanding the algebraic notation system is limited by poorly developed metalinguistic awareness, as we suggest, what can be done to assist the students concerned? Studying a traditional algebra course cannot be in their best interests. Research studies into the effectiveness of new approaches to algebra and alternative representational systems should be focused on those students. Teachers need to know whether a route can be found to help such students overcome the barrier of the notation system and gain access to the realm of important algebraic ideas.

REFERENCES

Brown, M. (1981). Number operations. In K. M. Hart (Ed.), *Children's understanding of mathematics: 11-16* (pp. 23–47). London: Murray.

Cuevas, G. J. (1984). Mathematics learning in English as a second language. *Journal for Research in Mathematics Education, 15,* 134–144.

Cummins, J. (1984). *Bilingualism and special education: Issues in assessment and pedagogy.* Clevedon, England: Multilingual Matters.

Herriman, M. (1991). Metalinguistic development. *Australian Journal of Reading, 14,* 326–338.

Mestre, J. P. (1988). The role of language comprehension in mathematics and problem solving. In R. R. Cocking & J. P. Mestre (Eds.), *Linguistic and cultural influences on learning mathematics* (pp. 201–220). Hillsdale, NJ: Erlbaum.

Secada, W. G. (1992). Race, ethnicity, social class, language, and achievement in mathematics. In D. A. Grouws (Ed.), *Handbook of research on mathematics teaching and learning* (pp. 623– 660). New York: Macmillan.

Stacey, K., & MacGregor, M. (1997). *Report of research project: Learning to use algebra for solving problems.* Unpublished report, Department of Science and Mathematics Education, University of Melbourne, Melbourne, Australia.

Tate, W. F. (1997). Race, ethnicity, SES, gender, and language proficiency trends in mathematics achievement: An update. *Journal for Research in Mathematics Education, 28,* 652–679.

Tate, W. F., & D'Ambrosio, B. S. (1997). Equity, mathematics reform, and research. *Journal for Research in Mathematics Education, 28,* 650–651.

APPENDIX

Test Used for Study II

1. Some of the following sentences can be written in a better way. If a sentence can be improved, write it in the better way.

    (i)   Red Riding Hood went to the forest with a basket.

    (ii)   While picking flowers, a wolf see her.

    (iii)   To hungrily drool it began.

2. Here is part of someone's bank statement, showing money deposited and withdrawn in January and February.

Date	Deposit	Withdrawal	Balance
08 JAN 91	120.00		120.00
12 JAN 91	30.00		150.00
23 JAN 91		50.00	100.00
03 FEB 91		40.00	
14 FEB 91		15.00	
26 FEB 91	120.00		

    Complete the balances for the last three dates.

    How many deposits were made during January?

    How much money was withdrawn during February?

3. In a class there are six more boys than girls.

    To find the number of girls, would you

    (i)   add 6 to the number of boys? OR

    (ii)   subtract 6 from the number of boys? (Choose one)

    If we write p for the number of girls and s for the number of boys, which of the following are correct?

    $s + 6 = p$ $p + 6 = s$ $s > p + 6$ $s - 6 = p$ $6p = s$ $6s = p$

4. Which of the following expressions can be written as $n + n + n + n + n$?

    $n + 5$ $n \times 5$ $5n$ n^5 5^n

Lessons Learned From Research

5. Explain the two different meanings of these two sentences:

   "He said he only liked me."   "He said he liked only me."

6. Put in brackets where necessary to make these sums correct.

   $4 \times 2 + 5 = 13$ $0.4 \times 2 + 5 = 2.8$

   $4 \times 2 + 5 = 28$ $0.4 \times 2 + 5 = 5.8$

7. Use algebra for the answers:

   (i)   David is 12 cm taller than Con. Con is b cm tall. David's height is _____.

   (ii) Sue weighs 5 kg less than Chris. Chris weighs y kg. Sue's weight is _____.

   (iii) Tess has twice as much money as Dino. Dino has $\$n$. Tess has _____.

8. The opening lines from the poem "Trapped Dingo" by Judith Wright are

   So here, twisted in steel, and spoiled with red

   your sunlight hide, smelling of death and fear,

   they crushed out of your throat the terrible song

   you sang in the dark ranges.

   Explain: (i) why the poem is called "Trapped Dingo" and

             (ii) the meaning of "spoiled with red your sunlight hide." _____.

9. If $9 - 4 = x - 9$, what number does x stand for? _____.

10. The distance from Melbourne to Townsville is 2487 km.

    You are drawing a map on which 1 cm represents 500 km.

    Which calculator keys would you press to work out the number of centimeters from Melbourne to Townsville on your map?

    (Circle one. Do not work out the answer.)

    $2487 + 500$ $2487 - 500$ $2487 \div 500$ 2487×500

11. Explain the meaning of the proverb

    "He who pays the piper calls the tune."

12. An operation $\otimes$ is defined by $a \otimes b = 2a + 3b$.

    Find x if $5 \otimes x = 22$.

13. Look at the numbers in this table and answer the questions.

x	0	1	2	3	4	5	6
y	2	5	8	11	14	17	...

(i) When x is 6, what is y?

(ii) When x is 10, what is y?

(iii) When x is 100, what is y?

(iv) Explain in words how you could work out y if you were told what x is.

(v) Use algebra to write a rule connecting x and y.

14. Which of the following numbers is the smallest?

A. 0.0908   B. 0.9008   C. 0.0098   D. 0.098   E. 0.908

15. $(0.3)^2 \times 0.8$ equals

A. 0.072   B. 0.0072   C. 0.72   D. 0.048   E. 0.48

16. Explain the meanings of this sentence:

"I sang yesterday as I wished."

17. By which number must $\frac{1}{2}$ be divided to obtain 3 as the result?

A. $\frac{1}{6}$ B. $\frac{1}{3}$ C. $1\frac{1}{2}$ D. 3   E. 6

18. Explain in your own words the meaning of the proverb

"Throw dirt enough and some will stick."

19. $7\frac{1}{2}\%$ of 200 is

A. 14   B. 15   C. 25   D. 18   E. 75

20. How many whole numbers are there between $\sqrt{50}$ and $\sqrt{500}$?

A. 14   B. 15   C. 62   D. 63   E. 449

Note. Difficult arithmetic items were included to challenge the more able students. Arithmetic scores were not used for the analysis described in this chapter. Maximum possible scores for language and for algebra were 14 each.

CHAPTER 14

DEVELOPING CONCEPTS OF SAMPLING FOR STATISTICAL LITERACY

Jane M. Watson and Jonathan B. Moritz, University of Tasmania, Australia

Abstract. A key element in developing ideas associated with statistical inference involves developing concepts of sampling. The characteristics of students' constructions of the concept of *sample* were studied through interviews of 62 students in Grades 3, 6, and 9. Six categories of understanding were identified; these categories should prove useful to teachers who want to help students develop appropriate understandings of how a sample must represent a population.

How do children develop the intuitions that form the foundation for sampling in statistics? Many educators seem to assume that students understand both the part-whole relationship of a sample to the whole population and the fact that the sample provides information, subject to variation, about what the whole population is like. Children, however, encounter examples of sampling in situations in which variation is not an issue; for example, they taste cola drinks or cheese in supermarkets. Teachers need to help children make the transition from these contexts to contexts in which representative sampling is required for making inferences. Coming to understand this transition is one of the educational goals underlying the research reported in this chapter. Another goal is for students to achieve, before they leave school, a level of statistical literacy that will allow them to contribute meaningfully to social decision making that is based on quantitative data. To achieve statistical literacy with respect to sampling, students must develop both understanding of sampling methods and ability to question claims that are based on biased sampling methods.

UNDERSTANDING ASPECTS OF THE CONCEPT OF SAMPLING

The importance of sampling in statistical analyses was stated explicitly in a general statement on statistical inference in *A National Statement on Mathematics for Australian Schools* (Australian Education Council, 1991):

The dual notions of sampling and of making inferences about populations, based on samples, are fundamental to prediction and decision making in many aspects of life. Students will need a great many experiences to enable them to understand principles underlying sampling and statistical inference and the important distinctions between a population and a sample, a parameter and an estimate. (p. 164)

Similar expectations can be found for students moving through the middle school years (Grades 5–8) in the United States.

The basic concept of sampling, in a context not involving variation, is noted by Moore (1991), who began his liberal arts textbook with "Boswell quotes Samuel Johnson as saying, 'You don't have to eat the whole ox to know that the meat is tough.' That is the essential idea of sampling: to gain information about the whole by examining only a part" (p. 4). In the context of a text focused on surveys, Orr (1995) introduced the term *population*, followed by a short section on sample:

This chapter is adapted from Watson, J. M., & Moritz, J. B. (2000). Developing concepts of sampling. *Journal for Research in Mathematics Education, 31,* 44–70.

This study was funded by Australian Research Council Grants A79231392 and A79800950.

Sometimes ... it is desirable to study (collect data from) less than the entire population. In such cases a subset of the population called a sample is selected ... [and] conclusions are drawn (generalized) to the larger population. ... What is the essential nature of a sample? In a word, a sample should be "representative," ... a small-scale replica of the population from which it is selected, in all respects that might affect the study conclusions. (p. 72)

The middle school book *Sample and Populations* (Lappan, Fey, Fitzgerald, Friel, & Phillips, 1998) includes a comprehensive discussion of issues associated with avoiding bias by choosing representative samples. Randomness and sample size, aspects of sampling considered important in this study, are included in the discussion.

APPLYING SAMPLING IN CONTEXT

Curriculum documents also state the need for students to be statistically literate participants in society. Thus, concepts taught in the statistics curriculum, including sampling, should be considered not only as basic concepts for data-collection projects in mathematics classes but also, in social contexts, as relevant concepts to be applied by students when they hear or read claims made in the media and elsewhere. This concern, including the need for students to adopt questioning attitudes, is beginning to be acknowledged in research focused on statistical literacy (Gal, 2000; Watson, 1997). The first author (Watson, 1997) has published a model for evaluating students' responses to questions involving statistical literacy. The model has three tiers in a hierarchy, the highest level of which represents the goal for students when they leave secondary school.

Tier 1

For this level, skills are associated with a basic understanding of statistical terms and topics, for example, *mean, chance,* and *graphing.* For sampling, Tier 1 concerns the concept of *a sample* as a small part that represents a population. For example, such a Tier 1 question as "What is a sample?" might evoke a response such as "A sample is a little part of something to show what it is like."

Tier 2

This level includes skills associated with recognizing and applying the basic language of statistics

in wider contexts. Often the context is social, but it may be scientific; such contexts abound in the media. Examples of Tier 2 understandings of sampling include recognizing that a survey reported before an election represents only a sample of the opinions of voters and being able to appreciate the meaning of confidence limits put on an estimate of voter support.

Tier 3

At this level, sophisticated skills allow a student to question statistical claims made without proper statistical foundation. In the preelection survey example, the sample size might not be reported or the sample size might be too small to allow for confidence to be placed in the estimate. Students operating in Tier 3 will be alert to such situations.

THE METHOD USED IN THIS STUDY

The students in this study, along with their entire classes, were administered written surveys during mathematics class time. Three survey items about sampling are shown in Figure 14.1.

Item Q1 was from a 20-item chance-and-data survey and was administered to all participants. Items Q2 and Q3 were from a media survey. Item Q2 was administered to Grade 6 and Grade 9 students, and Item Q3 was administered to Grade 9 students only.

For this study, we interviewed 62 students who had been surveyed: 21 from Grade 3, 21 from Grade 6, and 20 from Grade 9 (ages 8–9, 11–12, and 14–15 years, respectively). At each grade level, students were selected, on the basis of responses to survey items across many statistics topics, to represent a range of levels of understandings and to include some students who gave unusual responses. Teachers confirmed that the students selected by the researchers were articulate and willing to be interviewed. One or the other of us interviewed individual students in a separate room for 45 minutes each during class time; all interviews were videotaped. The interview protocol on sampling is shown in Figure 14.2. Eight other protocols used in the interview concerned other topics from the chance-and-data curriculum.

Analysis of Students' Answers

Our analysis of students' responses to all parts of the interview protocol and written-survey ques-

Q1. If you were given a "sample," what would you have?

Q2. ABOUT six in 10 United States high school students say they could get a handgun if they wanted one, a third of them within an hour, a survey shows. The poll of 2508 junior and senior high school students in Chicago also found 15 per cent had actually carried a handgun within the past 30 days, with 4 per cent taking one to school. (*The Mercury,* 21 July 1993, p.17)

(a) Would you make any criticisms of the claims in this article?

(b) If you were a high school teacher, would this report make you refuse a job offer somewhere else in the United States, say Colorado or Arizona? Why or why not?

Q3.

Decriminalise drug use: poll

SOME 96 percent of callers to youth radio station Triple J have said marijuana use should be decriminalised in Australia.

The phone-in listener poll, which closed yesterday, showed 9924—out of the 10,000-plus callers—favoured decriminalisation, the station said.

Only 389 believed possession of the drug should remain a criminal offence.

Many callers stressed they did not smoke marijuana but still believed in decriminalising its use, a Triple J statement said.

(*The Mercury,* 26 September 1992, p. 3)

(a) What was the size of the sample in this article?

(b) Is the sample reported here a reliable way of finding out public support for the decriminalisation of marijuana? Why or why not?

Figure 14.1. Sampling items from written surveys.

1. (a) Have you heard of the word *sample* before? Where? What does it mean?

(b) A news person on TV says, "In a research study on the weight of Tasmanian Grade 5 children, some researchers interviewed a *sample* of Grade 5 children in Tasmania." What does the word *sample* mean in this sentence?

2. (a) Why do you think the researchers used a *sample* of Grade 5 children, instead of studying all the Grade 5 children in Tasmania?

(b) Do you think they used a sample of about 10 children? Why or why not? How many children should they choose for their sample? Why?

(c) How should they choose the children for their sample? Why?

3. The researchers went to 2 schools:
   1 school in the centre of the city, and 1 school in the country.
   Each school had about half girls and half boys.
   The researchers took a random sample from each school:
   50 children from the city school,
   20 children from the country school.
   One of these samples was unusual: It had more than 80% boys.
   Is it more likely to have come from
   ☐ the large sample of 50 from the city school, or
   ☐ the small sample of 20 from the country school, or
   ☐ are both samples equally likely to have been the unusual sample?
   Please explain your answer.

Figure 14.2. Three parts of the interview protocol for sampling.

contexts: by identifying potential bias in the selection of children for the sample in a study of weight (Interview Part 2), by identifying the small sample from a school as more likely to be unusual (Interview Part 3), by questioning the sample from Chicago in the claim about gun use in the United States (Survey Q2), and by making one of several appropriate criticisms of the phone-in poll (Survey Q3) (e.g., "only youth listen," "only interested people call," or "callers may be lying").

RESULTS OF THE STUDY

The distribution of responses for the three grade levels over the six categories is shown in Table 14.1. Of the six categories of response, the first five categories reflected increasing use of appropriate mathematical content, ranging from limited or no experience with samples to a demonstrated ability to question claims. The students in the Equivocal Sampler category offered responses that could not be assigned to one of the first five categories; these students are discussed separately. In the first five categories, all Grade 3 students were *small* samplers, whereas all Grade 9 students were *large* samplers, with Grade 6 students offering the

tions led us to distinguish six categories of developing concepts of sampling. Most students offered quite similar descriptions of a sample; however, they differed in relation to ideas of adequate size, methods of selection, and recognition of bias.

In Part 2 of the interview, the students who suggested a sample of fewer than 15 were classified as *small* samplers, and those who said that the sample should contain 20 or more were classified as *large* samplers. No students suggested a sample size between 15 and 20. Methods of selection were associated with predetermination of the characteristic (e.g., "pick some fat and some thin"), with random choice, or with a primitive stratification procedure involving distribution from different schools (e.g., "pick some from each school in the state"). The students could demonstrate recognition of bias in four

Developing Concepts of Sampling for Statistical Literacy

Table 14.1

Number of Respondents by Category of Developing Concepts of Sampling and by Grade

Category	Grade 3	Grade 6	Grade 9
Small Samplers Without Selection	12	0	0
Small Samplers With Primitive Random Selection	4	1	0
Small Samplers With Preselection of Results	5	6	0
Large Samplers With Random or Distributed Selection	0	9	6
Large Samplers Sensitive to Bias	0	1	11
Equivocal Samplers	0	4	3

most diverse range of responses. We next briefly describe typical responses from each of the six categories. Italicized text within quoted extracts indicates the questions from the interviewers or written questionnaires.

Small *Samplers Without Selection*

Twelve Grade 3 students offered responses indicating that they were developing a basic concept of sample without any clear consideration for size or selection. In describing a sample, most of these students identified an example from their experiences and described a sample as "a small bit." Typical responses to the meaning of sample were "Like something free, a little packet or something" and "At [the supermarket] you get to try some food, and they put it in a little container for you so that you can try it." When asked how many children should be chosen in a sample for a study of the weight of Grade 5 children, all chose a size of fewer than 15. Only 5 of the 12 students offered any ideas for selection, and these were idiosyncratic ideas based on their experience of the context of children in the classroom, for example, "The teacher might just choose people who've been working well or something." Overall responses were remarkably uniform and limited to ideas about *sample* in the first tier of the statistical-literacy hierarchy.

Small *Samplers With Primitive Random Selection*

Five students who offered only partial descriptions in defining a sample and selected small sample sizes went on to suggest an early-developing concept of random or stratified selection methods. Three of these defined a sample in terms of a test, and two defined it as "a small bit"; no students affirmed both a test and a bit. All offered an example, again often of a food product or a product from a medical or a science setting. Although from the context of schools or Grade 5 classes these students had formed some isolated ideas about choos-

ing or had heard the word *random*, they could not explain the meanings of those ideas or relate them adequately to the context. The following excerpt is from the interview of one Grade 3 student.

Part 1b: Meaning in context? Some of the people. *Part 2a: Why not all?* Because they didn't have enough time. *Part 2b: How many? 10?* Umm, could have used 10. *Why?* It's an even number, and I like that number. *Part 2c: How choose?* Go by random. *Why?* Because they're not really worried about what people they pick.

Although these students' ideas about sampling went beyond those of students in the first category, they would still be considered within the first tier of statistical literacy; the students were still developing terminology. In applying sampling concepts in context, one must acknowledge the possibility of variation in the population and thus the importance of sample size and selection for representativeness. Students in the first two categories gave no indication that they appreciated variability; all their examples were of substances expected to have high uniformity.

Small *Samplers With Preselection of Results*

Acknowledging weight variation in the population, 11 students suggested selecting a sample of children by their weights, either to ensure that both "skinny" and "fat" children were in the sample or to exclude this variability and select only children who appeared to be about the normal weight. These students demonstrated that they had developed a basic concept of a sample as a small part to show what the whole is like, in that all 11 students included both "a small bit" and "a test or try" somewhere in their descriptions. In the context of studying weights, however, these students failed to realize that selection by weights would predetermine the results and, hence, would compromise the purpose of a sample: to see what the whole is like. The following Grade 3 student's responses illustrate these features. The

Lessons Learned From Research

idea of testing, initially of dirt, was later applied in the context of weights as a purpose for comparison, perhaps involving the idea of average.

Interview Part 1a: Meaning? . . . like if you were studying dirt, you'd find out just what was in it, if there were toxics or not. *Part 1b: Meaning in context?* Take out a few of them, I'd say. Not many. *Part 2b: How many? 10?* They could use any number, I'd say. Ten would be the . . . number I'd use. *Part 2c: How choose?* I would choose them in all shapes and sizes, some skinny, some fat. Then I'd compare them to another group and see what was the most average.

Although students in this category also chose small sample sizes and offered examples of samples of apparently homogeneous substances, they appeared to be moving into the second tier of the statistical literacy hierarchy in taking on a notable aspect of the context, variation.

Large *Samplers With Random or Distributed Selection*

Fifteen students demonstrated an understanding of the need for appropriately selected larger samples, but they did not mention potential bias in sampling. The two large-sample categories were different from the small-sample categories in respects other than size. All *large* samplers offered methods of selection, either by random methods or by selection from schools in different locations. Their descriptions of a sample were comprehensive, and many students also included a notion of average. The following responses from a Grade 9 student demonstrated awareness of the need for using a sample of a larger size to determine the average weight for the population as well as a method that involved selection from each school.

Interview Part 2b: How many? 10? They would need more than that to sort of find out what sort of average weight is for Grade 5 students. *How many?* 100, about. *Part 2c: How choose?* Pick a few schools, then just have 5 out of each school, 10 out of each Grade 5 so it's just sort of evened out . . . not all from one school, but a few different schools with only a few from each Grade 5 class.

The students in this category thus seemed to be consolidating concepts of size and selection of samples applied in context, concepts that are within the second tier of the statistical-literacy hierarchy.

Large *Samplers Sensitive to Bias*

Students in this category included all the features of students in the previously described category, but they exhibited the additional feature of recognizing that for the sample to represent the population, bias must be avoided. Twelve students offered responses demonstrating a comprehensive understanding of examples and defining characteristics of a sample: the need for a large size and for random or distributed selection methods, together with recognition, in at least two settings, of the potential for bias in sampling. These students often linked these ideas to the purpose of sampling as representing the whole or to the average as a representative measure. Although some still appeared to struggle to construct and apply their representative concept of sample in all settings, others consistently applied their sense of bias to question claims in various contexts. In the following responses from the only Grade 6 student in this category, the student failed to recognize bias in response to the first question about the Chicago sample (Survey Q2a) but acknowledged the variation in location when prompted (Q2b). When asked about sample size (Interview Part 2b), she also noted potential bias in the sample of weights.

Survey Q2b: Would you refuse a job somewhere else? The report would not make me refuse a job in Colorado or Arizona, but I would not take a job in a more dangerous place like Chicago or L.A. *Interview Part 2b: How many? 10?* Probably some more, because if they only used 10, they could all be ones that weighed about the same, and there could be some that weighed less and weighed more in other places. *How many?* Probably about 100 or something. *Part 2c: How choose?* Just choose anybody; just close your eyes and pick them or something.

Students choosing large sample sizes often used the word *random* and explained it carefully.

Interview Part 2c: How choose? Just randomly; you don't want to look at them; you just want to get a computer screen in the office . . . just randomly choose them all, just to make it fair. If you're going to choose all the fat children, then it's going to put the average right up, isn't it? And if you want to be fair. . . .

For others, a notion of a cross-sectional method of selection and the potential for bias permeated their responses, as follows.

Survey Q2a: Any criticisms? No, not really, because it is probably true, but the poll should have taken a larger cross section. *Survey Q3b: Is the sample reliable?* No, because a lot of people would not be bothered ringing in. Plus only the really active supporters would, due to the law being against them. *Interview Part 1a: Meaning?* … in surveys and stuff … a sample is like a cross section of a community or whatever, or like at the supermarket—they give away samples like a piece to try. … *Part 1b: Meaning in context?* They interviewed, like, a cross section, all different students from here, there, and everywhere all round Tasmania. They didn't like use every single student, … just randomly picking them. *Part 2b: How many?* 10? I think they would use a lot more than 10 because if they'd used 10, it could be a lot under or a lot above because you could have extremely high results. … *How many?* It depends how many Grade 5 children there are in the state, perhaps 10% or 20% of Grade 5 students in the state. *Part 3: Larger or smaller sample for unusual result?* I think it would be more likely to get it [the sample of 80% or more boys] from the country school because there are a lot less children, so if you had perhaps a few more, you would bring the percentage up a lot quicker than with this sample [city school].

The understanding displayed in this response reflected the third tier of statistical literacy in terms of a consistent appreciation of characteristics of good samples across contexts and ready recognition of potential bias.

Equivocal Samplers

Seven students offered responses that were in some sense equivocal in terms of the choice of sample size. We were therefore unable to assign their responses to one of the five categories described above. For several students, an apparently tentative choice associated with one of the *small-* or *large-*sampler categories was mitigated by a method of selection that was more typical of the other group. In one response, the dominant idea was that sample size did not matter, although the student suggested 20 or 30 "so that's just like a sample." Another student was more concerned about divisibility of the sample size into the total and hence did not appreciate the underlying issue of sample size. A third

student did not resolve conflicting ideas about representativeness and variability of samples. He responded not only that any number was appropriate for a sample size but also that the whole population should be tested to be accurate. Two Grade 9 students suggested random methods of selection and thought that large sample size was important but that a size of 10 or 14 was sufficiently large. Equivocal Samplers appeared to be in transition between the *small* samplers and the *large* samplers and displayed understanding typical of students in the second tier of statistical literacy.

DISCUSSION

Categories of Developing Concepts of Sampling

The hierarchy of categories observed in students' responses may be hypothesized as a model of students' development of concepts of sampling. Students initially build a concept of sample from experiences with sample products or in medical- and science-related contexts, perhaps associating the term *random* with sampling. When students begin to acknowledge variation in the population, they recognize the importance of sample selection, at first attempting to ensure representation by predetermined selection but subsequently by realizing that adequate sample size coupled with random or stratified selection is a valid method to obtain samples representing the whole population. When valid methods of sampling are consolidated, sample data are interpreted with appreciation of how sample size and selection contribute to biased or representative samples.

The six categories of response can be seen to fit within the three tiers of the hierarchical model for statistical literacy, introduced earlier.

Tier 1: *Small* Samplers Without Selection
    Small Samplers With Primitive Random Selection
Tier 2: *Small* Samplers With Preselection of Results
    Equivocal Samplers
    Large Samplers With Random or Distributed Selection
Tier 3: *Large* Samplers Sensitive to Bias

Tier 1 students are developing a concept of sample. They have a basic idea of the language, often expressed in ideas related to sample, and sometimes a context-free use of the idea of randomness.

Tier 2 students have richer concepts of sample that can be applied in straightforward contexts, although some have not yet recognized appropriate selection techniques and some struggle with sample size. Tier 2 encompasses the largest range of ideas: some students retaining small-sampling ideas but within a context in which the importance of sampling methodology is recognized, some apparently in transition with conflicting ideas unresolved, and some with well-developed sampling and selection methods but unable to go to the critical stage. The sense of sample is well developed for Tier 3 students, who understand appropriate selection methods and recognize bias in many or all situations in which it can occur. Except for several students in the Equivocal Samplers group, students generally did not criticize biased samples without also exhibiting appropriate understandings of sample size and selection. The results support the hypothesis that the development of concepts of sampling progresses hierarchically.

This framework for describing developing concepts of sampling may be useful to teachers and curriculum planners who are interested in providing experiences to help students structure responses that are more complex and at the same time achieve higher levels of statistical literacy in social contexts. Two key concepts emerge as significant for developing concepts of sampling. One is the role of the appreciation of population variation in the transition from Tier 1 to Tier 2: Students recognize the need for larger sample sizes and also for random or stratified selection methods. Consideration of variability in populations thus deserves explicit attention in statistics curricula. The other key concept is consistent sensitivity to bias in the transition from Tier 2 to Tier 3. A significant result of this study is that a student's demonstration of a preference for large random samples is not sufficient to ensure that the student will identify sample bias, as evidenced by the 15 *Large* Samplers With Random or Distributed Selection who did not acknowledge bias. Even some *Large* Samplers Sensitive to Bias did not notice bias in all contexts. Students' success in identifying bias may be heavily dependent on contextual clues.

Teaching Implications for Developing Concepts of Sampling

As with other terms in mathematics, the term *sample* has a relatively straightforward meaning when met in out-of-school contexts and a more technical meaning that applies when one uses it to represent a population in an unbiased fashion in order to draw conclusions about that population. The transition from the former meaning to the latter involves reconstructing the notion that suffices in situations such as receiving a sample product at a supermarket, where appreciable variation is unlikely and a small sample is adequate. Such situations do not help to generate appreciation for the variation present in many wholes or populations nor for the consequent need for samples to be large enough and representative enough to show realistically what the whole is like. Explicit discussion about collecting sample data for which the measure, such as height or weight, obviously varies in the population may be important for developing concepts of sampling.

As well as teaching appropriate methods for selecting samples, teachers must help students develop appreciation for situations in which bias can occur. In this study most Grade 9 students appreciated the need for larger samples and appropriate methods of selection but often did not produce convincing arguments based on representativeness and avoiding bias. Teachers risk seeing no improvement in this situation if they implicitly assume that students who favor random sampling will also appreciate how it is applied in real contexts or if they encourage students to focus merely on sample size reported in statistical claims, without identifying bias in the sampling procedures in the context. We suggest that at all levels, the need to use representative samples requires reinforcement. In magazines popular with teenagers, the use of voluntary-response surveys to make claims about the magazine's readership or wider populations seems to be a good starting point to motivate discussion. Using examples from students' personal experiences and the media should help motivate the questioning attitudes required of future citizens.

REFERENCES

Australian Education Council. (1991). *A national statement on mathematics for Australian schools*. Carlton, Australia: Author.

Gal, I. (2000). Statistical literacy: Conceptual and instructional issues. In D. Coben, J. O'Donoghue, & G. E. Fitzsimons (Eds.), *Perspectives on adults learning mathematics* (pp. 135–150). Dordrecht, The Netherlands: Kluwer.

Lappan, G., Fey, J. T., Fitzgerald, W. M., Friel, S. N., & Phillips, E. D. (1998). *Samples and populations* (Teacher's Ed.). Menlo Park, CA: Dale Seymour.

Moore, D. S. (1991). *Statistics: Concepts and controversies* (3rd ed.). New York: Freeman.

Orr, D. B. (1995). *Fundamentals of applied statistics and surveys*. New York: Chapman & Hall.

Watson, J. M. (1997). Assessing statistical thinking using the media. In I. Gal & J. B. Garfield (Eds.), *The assessment challenge in statistics education* (pp. 107–121). Amsterdam: IOS Press and The International Statistical Institute.

CHAPTER 15

WHEN A STUDENT PERPETUALLY STRUGGLES

Kristine K. Montis, Minnesota State University—Moorhead

Abstract. Dyscalculia is a psychological and medical term that refers to extreme difficulty in learning mathematics and, in particular, to deficits in the production of accurate, efficient arithmetic calculations. A relationship between difficulties in language processing and difficulties in mathematics learning was examined in this yearlong, qualitative case study of a 12-year-old student who displayed many characteristics of dyscalculia. The student's learning experiences during her school mathematics and tutoring sessions demonstrated the vital role language processes play in developing the concept flexibility necessary for success in mathematics. The study has implications for pedagogy in classrooms that include mainstreamed students with learning disabilities.

OCCASIONALLY a teacher of mathematics encounters a student who, although willing and hardworking, seems to perpetually struggle and often fail in mathematics class. This failure is especially frustrating, disappointing, and baffling when other students have been successful with the class lessons, activities, and discussions. Such a struggling student, even working one-on-one and using supportive, concrete representations for the mathematics to be learned, often still "just doesn't get it." So what makes mathematics extremely difficult for such a student, and what else can we as teachers do to support and nurture mathematical learning in such students?

This question has no easy or universal answers, but researchers are beginning to suspect that, for some of these students, receptive language and phonological-processing problems are the underlying cause of mathematics-learning difficulties. This chapter summarizes the findings from a case study of a student who exhibited many characteristics associated with *developmental dyscalculia*, a medical term referring to extreme mathematics-learning difficulty not related to traumatic brain injury. I describe observations that led to the conclusion that the student's learning difficulties were phonological in nature, and, finally, I discuss ideas and recommendations for teachers faced with students having similar difficulties.

During Kay's fifth-grade year, I worked with her as a tutor and researcher. Kay was the type of student about whom I, as a teacher, had always worried and thought that I should be able to do more to help. She attended school regularly, was cooperative, and diligently attempted all her schoolwork. But in mathematics and reading, she continually struggled and often failed, even with the one-on-one support I was providing. Kay's classroom teacher agreed that Kay tried hard and wanted to learn but said that Kay's responses were sometimes "just out of nowhere and made no sense." This observation finally led to a breakthrough in my work with Kay and to my understanding how phonological-processing deficits directly affected mathematics learning in Kay's case.

Phonological processing refers to the brain's processing of speech sounds. Children suffering from deficits in phonological processing are hypothesized either to compensate by pulling meaning from context or to struggle through life, living in a language fog (Blakeslee, 1995; Lyon 1995). Phonological-processing deficits do not show up in

This chapter is adapted from Montis, K. K. (2000). Language development and concept flexibility in dyscalculia: A case study. *Journal for Research in Mathematics Education, 31*, 541–556.

standard school hearing tests because the student is able to hear the sound but is unable to accurately process or interpret the sound. A related term, *phonemic difficulties*, is found in research on reading difficulties. *Phonemic difficulties* are difficulties in recognizing or differentiating among speech sounds as they are used in language and are suggested to be the basis of difficulties in learning to read (Siegel, Share, & Geva, 1995; Snow, Burns, & Griffin, 1998). From an information-processing point of view, this type of difficulty also results from a deficit in phonological processing.

An example of how the "fuzzy" phonological perception affected Kay is the "red colt" incident reported by her teacher. In this instance, Kay's teacher had read the class a story in which George Washington borrowed a horse when the redcoats were coming. In her answers about the story, Kay pulled meaning from context and answered questions about the story by referring to a "red colt" that was not actually in the story. I believe that in trying to make sense of what she had heard, Kay guessed "red colt" instead of "redcoat" in the presence of a cue that a horse was involved.

Although the red-colt incident did not occur in a mathematics class, I believe that it illuminates, by analogy, the several incidents in mathematics tutoring sessions in which the difficulty might be attributed to a phonological-processing deficit. I had already noticed that Kay often confused such similar-sounding terms as *five, fifteen, fifty,* and *fifths.* However, for several months I failed to understand the extent of the difficulties this fuzzy perception was causing Kay. The following episodes describe how I concluded that the phonological deficit was tied directly to Kay's difficulty in developing facility with multiple representations (object, picture, spoken word, written symbol) of fractions, as well as in developing the concept of fraction equivalence.

As part of her learning to name fractions, Kay and I made models of fractions by using colored index cards. We made halves, thirds, fourths, and eighths, each in a different color. We practiced making fractions, naming them verbally and in writing. Then we struggled to develop Kay's concept of equivalent fractions.

Only through extensive questioning, reflection, and analysis did I realize that Kay could relate fractions to the written symbols and could correctly fill in the blanks indicated in Figure 15.1 but that she

had not yet grasped the verbalization of the denominator as ending with *th.* For the first picture, therefore, Kay answered "three" or sometimes "three over four" but not "three fourths," and for the second picture, she answered "four" or sometimes "four over eight" but not "four eighths." Her fuzzy perception of the terms *eight* and *eighths* had led her to construct the fraction as "four pieces shaded over eight pieces total." She had missed the phonological cue provided by the *ths* in the standard fraction-naming scheme.

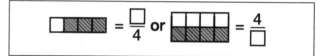

Figure 15.1. Fraction representations not associated with verbalization of denominator as ending in *th.*

The phonological cue of *ths* is essential both in understanding the fractional nature of the amount being named and in recognizing the existence of a fractional relationship between the numerator and the denominator. To Kay, this relationship was just the positional relation *over* between two whole numbers. She was not perceiving the *th* cue or the denoted fractional relationship even though she appeared to be saying and writing the correct answers. Because she had missed the phonological cue provided by the *ths* in the standard fraction-naming scheme, Kay had constructed the numerator and the denominator as separate entities with no connecting relationship. Therefore, because she did not perceive the auditory cue, she had never tuned in to the idea of a fraction as representing a relationship and was unable to proceed to the conceptualization of equivalent fractions as two representations of the *same relationship.* Once the *th* language denoting the relationship of numerator to denominator was pointed out to Kay, she quickly learned to recognize and name equivalent fractions. Almost immediately, she was able to correctly apply the procedure she had been taught in the classroom for finding equivalent fractions, whereas she had previously been applying it in a haphazard and often incorrect fashion.

When improper fractions were introduced in her classroom, we again used the index-card models. Kay could model five halves as five of the one-half pieces of index card and then, after working on the concept, could also show five halves as two wholes and an extra half. But when I encouraged her to relate this thinking to her homework on mixed numbers and

Lessons Learned From Research

improper fractions, she did not spontaneously make the connection between the written symbols and our activities. Her conceptual constructions were inflexible, attached to a specific physical model or representation. She was unable to connect the idea of an index-card model called *five halves* with a written symbol (5/2) that was called *five halves*. For most people, the verbalization *five halves* provides a transitive bridge between the model and the written symbol. Many students must have this relationship pointed out at first; once it has been so noted, most students readily develop the concept flexibility necessary to use any of the concept representations experienced up to that point.

But for a student with a processing disorder like Kay's, the difficulty in making these connections is magnified by her fuzzy perception of speech sounds. She not only missed such phonological cues of distinction as the *ths* in fractions but also ignored phonological cues of generalization, such as naming two different representations with the same word name. In Kay's world of fuzzy perception, she had learned not to trust that vocalizations that sounded the same did, in fact, represent the same things. Therefore, Kay had not developed an understanding of the language of the transitive property that *if a equals b, and b equals c, then a equals c*. In this way, Kay's phonological-processing difficulty complicated for her the process of establishing meaningful connections among the physical models and the verbal and written symbols because she could not trust her perception of similarity in spoken words. Instead, when Kay constructed meaning for each new representation, for instance, for the written symbol 5/2, she also needed to specifically connect this new representation to every other representation that she had previously experienced. Kay was not able to spontaneously make compound or transitive connections among the representations on the basis of a common vocalization of each quantity as "five halves."

CONCLUSION

The evidence collected in this case study showed that Kay was strong in visual organization and used context cues extensively to make sense of her world, including her mathematics experiences. She used a quiet demeanor to prevent others from noticing her frequent confusion and projected an often-misleading appearance of comprehending verbal instruction. Receptive language was a substantial area of weakness and was the basis of her learning difficulties. However, this weakness was not readily apparent because of her ability to respond appropriately, at least superficially, by using context clues.

This case demonstrates how one student's learning difficulties in mathematics may be related to phonological-processing deficits similar to those now being suspected in many reading disabilities (Adams & Bruck, 1995; Blakeslee, 1994; Lyon, 1995; Risey & Briner, 1990). The inability to clearly hear or process the phonemic components of language has obvious implications for developing language and reading skills. The connection to mathematics-learning difficulties becomes apparent when a phonological-processing deficit is introduced to transitive thinking processes used to develop mathematical concepts and representations. For instance, if a model is described in words as *three fourths* and the written symbol is read as *three fourths*, the student with a phonological-processing deficit has no way of knowing whether she should regard these terms as the same. What for most students is a natural development of connections among ideas, terms, and representations based on a transitive use of language is problematic and frustrating for a student like Kay. The situation is also frustrating and baffling to the teacher who is not aware of this type of learning difficulty. Compounding the problem is the likelihood that the deficit may remain unidentified, as it did with Kay, even if standard diagnostic testing is done. This deficit-detection difficulty may be a major reason for the rarity in diagnosis of mathematics disabilities and the intransigence of the belief that some people just cannot do mathematics.

The first step in remedying this situation is for teachers to become aware of the possibility of phonological-processing deficits as a source of students' learning difficulties in mathematics. Characteristics indicative of a phonological-processing deficit include (a) unusual difficulty in learning mathematics or in reading; (b) especially quiet, watchful behavior; (c) reticence to speak in class; (d) odd or incongruent responses to information received verbally; (e) sound-alike word substitutions; (f) apparent confusion or uncertainty when asked simple, informational questions; (g) physical symptoms, such as finger agnosia (inability, with eyes closed, to tell which finger has been lightly touched), imbalanced body posture, left-right confusion; and (h) family history of epilepsy or Attention Deficit (with Hyperactivity) Disorder (ADHD). If a student has several of these characteristics, a hypothesis of phonological deficit should be considered.

Second, teachers should use caution when considering existing diagnoses of student's learning difficulties. As in the case of Kay, legal labels and diagnoses are often inadequate and misleading (Lyon, 1995). The dilemma of the classroom teacher lies in interpreting and applying such diagnoses or the absence thereof. According to the legal definition of *learning disability*, Kay, at the beginning of the study, was not considered a learning-disabled student but instead was categorized as a slow learner, because her measured IQ was below the legal cutoff for learning disabilities. Technically, she was over-achieving for her IQ. According to that categorization, she could not be expected to improve her performance and needed only more time, not special materials or approaches, to achieve what other students achieved more quickly and easily.

Both assumptions interfere with expectations of and efforts to assist such a student. Kay was, in fact, able to improve her achievement in both reading and mathematics. Although more time was necessary for Kay's continued growth, another essential factor was the realization that the use of transitive language and thinking was problematic for Kay. This situation required that the instructor take the initiative to understand how Kay perceived and understood the world. The challenge for the instructor is to understand and cope with a student for whom transitive language connections are not sensible and automatic. Clearly, the teacher must bridge this gap, especially because such students, out of embarrassment, commonly attempt to hide their confusion instead of attempting to seek clarification.

Another implication for classroom instruction has to do with accommodations commonly needed to support students with cognitive-processing deficits. Because cognitive-processing deficits require affected students to work more slowly and expend greater effort, those students need extended time to do assignments and close supervision of independent-work periods. The role of the supervisor in this setting is to help the student stay on task, monitor the student's frustration and fatigue levels, and adjust the assignment length or process accordingly. In a classroom with a teacher's aide, the aide may be trained to serve in this role.

The final pedagogical implication of this study is that learning mathematics cannot reasonably be divorced from learning language and making sense of the world as a whole. Over the years, many theorists have noted these connections, but educators have yet to develop adequately sensitive methods for recognizing and dealing with deficits in language development in relation to learning mathematics. Because of fuzzy perception and processing interference, a student with such disabilities has difficulty adequately organizing experience into shared knowledge. Such students have an additional problem learning to see and interpret the world the way other people do so that they can, in turn, develop language with which they may accurately and fluently operate with the ideas and abstractions of mathematics. The challenge this problem presents for mathematics educators is immense. Through this case study, mathematics educators can realize that these types of difficulties exist; understand how they may be manifest in mathematics classrooms; and learn ways that student achievement, understanding, and confidence may be facilitated.

REFERENCES

Adams, M. J., & Bruck, M. (1995). Resolving the "great debate." *American Educator*, *19*(2), 7, 10–20.

Blakeslee, S. (1994, August 16). New clue to cause of dyslexia seen in mishearing of fast sounds. *New York Times*, pp. C1, C10.

Blakeslee, S. (1995, November 14). "Glasses for the ears" easing children's language woes. *New York Times*, p. C1.

Lyon, G. R. (1995). Research initiatives in learning disabilities: Contributions from scientists supported by the National Institute of Child Health and Human Development. *Journal of Child Neurology*, *10* (Suppl. 1), s120–s126.

Risey, J., & Briner, W. (1990). Dyscalculia in patients with vertigo. *Journal of Vestibular Research*, *1*, 31–37.

Siegel, L. S., Share, D., & Geva, E. (1995). Evidence for superior orthographic skills in dyslexics. *Psychological Science*, *6*, 250–254.

Snow, C. E., Burns, M. S., & Griffin, P. (Eds.). (1998). *Preventing reading difficulties in young children*. Washington, DC: National Academy Press.

SECTION III

RESEARCH RELATED TO CURRICULUM: INTRODUCTION

THE curriculum studies included here are varied. Some authors compare already-developed curricula, whereas others describe the development and evaluation of new curricula. Some of this work was done in only one classroom, and other studies took place in several classrooms. Excellent tasks used in some studies are included here.

Chapter 16. "Open and Closed Mathematics: Student Experiences and Understandings," by Jo Boaler

This researcher spent 3 years studying two secondary schools in which very different approaches were taken to teaching mathematics. One school used a traditional approach, and the other used an open-ended approach. The researcher was particularly interested in whether different forms of learning would take place in the two schools and whether students would react differently in new or unusual settings. Serendipity played an important role in the author's having the opportunity to conduct this research study, which took place in England when a new type of national mathematics examination was being pilot tested. Because administrators and teachers at one school knew that their students would not have to take the regular examination, they took advantage of the school's commitment to progressive education to experiment with their mathematics curriculum. The researcher collected many types of data, including interviews, questionnaires, observation notes, and assessment data, over the 3 years. From her analyses of these data, she found that students in the school with an open approach to mathematics performed much better on tasks that required that they apply their knowledge to solve unfamiliar tasks. She also found that the students at both schools performed equal-ly well on traditional, closed test items. The many student comments throughout the chapter are informative because they tell the reader what students believed to be the nature of school mathematics. The differences in comments from the two schools are striking.

Lessons learned

Two very different curricula were compared in this study. Powerful evidence of successful learning in informal settings is presented in this chapter. Some results are predictable, but some are surprising—particularly the finding that students who experienced an open-ended approach to mathematics performed as well on traditional tests as did students who followed a more traditional curriculum.

The two approaches to mathematics represent very different value systems. Teachers at the progressive school were most likely teaching all subjects in nontraditional ways. This approach is attractive to many parents and teachers, but other parents and teachers are opposed to this approach to schooling. The choices of approaches and curricula and the ways in which these choices represent value systems extends well beyond the location of this study (England) and the grade levels involved (secondary). All teachers will find this chapter provocative. Many administrators, parents, and politicians will also find this chapter interesting.

Chapter 17. "Developing Children's Understanding of the Rational Numbers: A New Model and an Experimental Curriculum," by Joan Moss and Robbie Case

The study described in this chapter represents a successful collaboration between an experienced teacher and a developmental psychologist. The psychologist applied his knowledge of the way children develop intellectually to conjecture that children would be more successful in learning rational numbers by beginning with percents, then moving to decimals, and only then to fractions. This sequence is probably the reverse of the way in which every reader of this chapter learned rational numbers.

The researchers worked together to develop a curriculum, infused with meaning of numbers and number operations, in which this sequence of instruction was followed. The curriculum was then taught to one class of fourth graders; a class using a traditional curriculum served as a control group. The results, based on preinterviews and postinterviews with each student in each class, are as predicted: The students in the experimental class outperformed the students in the control class. Many responses given by students from the two groups are included, and they are particularly revealing of the depth of knowledge acquired by the students in the experimental class.

Lessons learned

Are you able to quickly and mentally calculate 65% of 160? The fourth graders using the experimental curriculum could do so. The lessons here are of two kinds. One is that the usual way in which educators sequence topics in the mathematics curriculum is not sacrosanct. Could instruction be sequenced differently in any other areas? Certainly the sequencing in the integrated curriculum of most countries is very different from the algebra-geometry-algebra curriculum more common in U.S. secondary schools. In the elementary and middle grades, should two-dimensional geometry precede or follow three-dimensional geometry? Or, as illustrated by Jacobson and Lehrer (Chapter 10), can we begin primary-grade geometry instruction with transformational geometry?

Another lesson to be learned from this study is the power of a developmentally appropriate, well-constructed, well-taught curriculum. The fourth graders in the experimental class acquired deep, rich, connected knowledge of rational numbers; they could use that knowledge to make sense of problems presented to them. This lesson applies well beyond instruction in rational numbers in fourth grade.

Chapter 18. "A Comparison of Integer Addition and Subtraction Problems Presented in American and Chinese Mathematics Textbooks," by Yeping Li

Educators have heard of the TIMSS studies in which Asian students outperformed U.S. students. This researcher examined five American textbooks and four Chinese textbooks to find similarities and differences in the ways in which mathematics instruction was approached. The mathematics examined was limited here to integer addition and subtraction. All the Chinese textbooks but none of the American textbooks embedded integer operations within units on rational numbers; thus the Chinese students were learning to add and subtract positive and negative rational numbers at the same time as they learned these operations on integers. More Chinese textbooks emphasized procedures, whereas more American textbooks emphasized conceptual understanding. These differences, however, were not large.

Lessons learned

What we learn from international studies depends on what the researchers look for and report to us. Perhaps tests do not tell the whole story.

Chapter 19. "Supporting Latino First Graders' Ten-Structured Thinking in Urban Classrooms," by Karen C. Fuson, Steven T. Smith, and Ana Maria Lo Cicero

Once again we consider a study in which a developmental model was used to construct a curriculum, this time for teaching two-digit addition and subtraction. The study took place over the course of a year in two predominantly Latino, low-socioeconomic classes, one taught in English and one taught in Spanish.

The model is detailed and comprehensive and thus will take some time for the reader to come to understand. With careful study, the drawing in Figure 19.1 will begin to make sense, particularly to teachers who have tried to teach place-value concepts to primary-grade children.

Figure 19.1 in this chapter shows, along the right side, various ways that children think about the two-digit number 53. For each of these ways, students must make connections among the quantity (the things being counted), the words we use to tell the value of the quantity, and the written numeral. The authors call these relationships *triads*. For example, a child at the unitary stage might count a set of objects beginning with 1 and ending with 53, without any understanding of what the 5 and the 3 represent. He uses the words *fifty-three*. He might also associate the symbol 53 with the words. This outer triad is shown in Figure 19.1 on the top right as 53 dots, on the far left as number words, and at the bottom right as the symbol 53. By following the outside dotted lines, one moves from the actual counting to the words to the symbol. This relationship is the unitary triad. Note here that when the authors used double arrows to make connections between elements of a triad (and a triad for each step of the sequence along the right side), they realized that they could not draw double arrows in every case because the quantities and written numerals were not always linked in children's conceptions.

When children first count using two-digit numbers, they do not distinguish between the ones and tens. They see all numbers as "unitary." When they begin to give meaning to each digit in a two-digit number, they move to what the authors call a *decade-and-ones conception* (follow the middle column of Figure 19.1). The researchers point out that all children move from the unitary conception to the decade-and-ones conception but that the next two conceptions are not developmental—whether a child uses a *sequence-and-ones* conception or a *separate-tens-and-ones* conception depends on instruction. Either can lead to the final *integrated tens and ones conception*. Thus, one major hurdle for these researchers was to decide on which instructional conceptions to focus in their curriculum. Only after careful consideration of the way numbers are named in various languages did they decide to focus instruction on the separate-tens-and-ones conception. For example, the Chinese would call 53 "five ten three" and thus need only the separate-tens-and-ones structure following the universal unitary conception of number.

This decision led to other decisions. The children were taught to say "one ten two ones" for 12 but needed also to become familiar with ordinary names (twelve in English, doce in Spanish). Another decision made was to use penny frames, in which pennies fit 10 in a row, because the frames were easy to manage, inexpensive, and required minimal teacher-preparation time. This decision illustrates the researchers' belief that a concrete aid is needed to teach place value but that it need not be a standard one, such as

base-ten blocks. Finally, the authors decided to use a particular scheme of recording the children's addition and subtraction work so that, after class, the teachers would have access to the records of the work.

Each time a curriculum is studied, researchers attempt to measure its success in some way, usually by comparing its results with those of another curriculum. These researchers decided to compare these first-grade students' performance with East Asian children's performance as reported in other research studies. This comparison showed that the first graders in the study performed more like the East Asian children than like children in higher grades in the United States.

Lessons learned

This curriculum was very successful in inner-city schools, and this fact in itself makes this study interesting. The developmental model used in this study is challenging, but it makes sense once it is understood—and pleasure is gained in coming to understand the model, just as in solving an interesting problem. The lessons to be learned include the manner in which instructional decisions were made. The researchers worked closely with the classroom teachers in this study, and the shared decision making helped make this study successful. The children's recording of operations using lines (sticks) and dots was credited with much of the success because it was an easy way for students to show their work and a tool to use during assessment. A major lesson is that one can develop a solid curriculum that builds on students' prior knowledge, is carefully designed, is not too difficult for teachers to use, and provides a way to record work. This work is a prototypical example that is worth testing in other classrooms, in other curricular areas, and at other levels.

Chapter 20. "Effects of Standards-Based Mathematics Education: A Study of the Core-Plus Mathematics Algebra and Functions Strand," by Mary Ann Huntley and Chris L. Rasmussen

Here is a classic study comparing two curricula—the type of study many parents are calling for in school districts considering adopting a "reform" curriculum, such as Core-Plus. This type of study is extremely difficult to carry out because to get meaningful results, researchers must (a) include fairly large numbers of students in the study, (b) randomly assign students to one of the two curricula or at least match them in appropriate ways, (c) control for the teacher factor, and (d) select or devise assessment measures fair to both curricula.

In various parts of the United States, the researchers located six schools, each of which had two teachers who taught the Core-Plus curriculum and one, two, or three control teachers who taught a more traditional curriculum. Students in Core-Plus and control classes either were of comparable ability or were randomly assigned to Core-Plus or control classes. Three assessments were designed— one that measured algebraic problem-solving ability, one that measured understanding, and one that measured manipulative skills.

The results were not surprising. The Core-Plus students exhibited better understanding of the content and better problem-solving skills. The students in a traditional curriculum were more successful at manipulating symbolic expressions used in algebra. However, none of the tests resulted in exemplary scores, and in that sense, they were disappointing.

Lessons learned

One lesson learned is also a classic one— Students will not learn what they have not had opportunities to learn. Core-Plus students had opportunities to learn mathematics with understanding and to solve algebraic problems, but they did not spend much time practicing skills. An important question of values is implicit in the results of this study. What do teachers and parents want for their students, for their children? Do they want them to understand mathematics, or is skill proficiency the hallmark of a good mathematics curriculum? Many educators would like to think that students in Core-Plus classes will be more likely than students in traditional classes to succeed in college-level mathematics. Many college-placement examinations, however, continue to focus on manipulative skills, and a student who lacks those skills may be held back to repeat high school level courses. The message many parents take from this situation is that skills are paramount and that the Core-Plus curriculum is not a good one.

Decisions regarding curriculum choices always come down to questions of values. Do we,

as educators and parents, value understanding, problem-solving ability, and believing in oneself as capable of doing mathematics? Or do we value skills because skills were the focus of our own mathematics programs and we believe that is why we are successful? Is knowing what any one curriculum can provide sufficient information for us to decide whether that curriculum is a good one, or do we need comparative research studies, such as this one?

Chapter 21. "Good Intentions Were Not Enough: Lower SES Students' Struggles to Learn Mathematics Through Problem Solving," by Sarah Theule Lubienski

This study is an interesting counterpart to the previous one in that the study begins with a tentative assumption that a problem-centered curriculum is good for everyone, including students having low socioeconomic status. The researcher explores this assumption within the context of the students' learning in her own classroom.

Few mathematics education researchers have considered social-class differences separated from differences of gender or ethnicity. The researcher in this case was also the teacher of a seventh-grade class in which she was pilot-testing a reform curriculum. Her students represented a mix of socioeconomic levels. Only 3 of the 30 were not Caucasian. Half were female. The researcher studied the differences she found in students' reactions to the curriculum. Briefly stated, she found that students having lower SES, in contrast with those having higher SES, typically preferred the "old math" in which little problem solving was required and in which procedures were clear, and they preferred more teacher direction than was appropriate for a teacher using the reform curriculum. They saw the benefits of the new curriculum only in terms of what they *should* be learning, whereas the students having higher SES spoke of the benefits in terms of how they were personally helped by the program. Each factor is discussed by the author, who also discusses implications of her study.

Lessons learned

This study raises more questions than it answers. One teacher-researcher in one classroom with a new curriculum cannot arrive at

Lessons Learned From Research

definitive answers to the questions she is asking in this study. But it is a start. This study presents a great deal of information for educators to consider. Why does socioeconomic status make a difference in how students learn and react to a particular curriculum? Is possible harm done in offering them a problem-solving curriculum? To what extent are the results of this study due to the students' experiences in the earlier six grades with a more traditional curriculum in which students having lower SES felt more successful? Will these results hold after students have experienced this or a similar curriculum over several years? We have ample need for more research to explore these questions.

OPEN AND CLOSED MATHEMATICS: STUDENT EXPERIENCES AND UNDERSTANDINGS

Jo Boaler, Stanford University

Abstract. Three-year case studies of 2 schools with different mathematics-teaching approaches were conducted. One school used a traditional, textbook approach; the other used open-ended activities at all times. Various forms of the data, including observations, questionnaires, interviews, and quantitative assessments, show the ways in which the 2 approaches encouraged different forms of learning. Students who followed a traditional approach developed procedural knowledge that was of limited use in unfamiliar situations. Students who learned mathematics in an open, project-based environment developed conceptual understanding that yielded advantages in a range of assessments and situations.

Agrowing concern within mathematics education is that many students are able to learn mathematics for 11 years or more but are then unable to use this mathematics in situations outside the classroom context. Various researchers (e.g., Lave 1988; Nunes, Schliemann, & Carraher, 1993) have shown that adults and students, when presented with "real world" mathematical situations, did not use school-learned mathematical methods or procedures but instead used their own invented methods.

Various mathematics educators have suggested that students are unable to use school-learned methods and rules because they do not fully understand them. Educators relate this lack of understanding to the way that mathematics is taught. Schoenfeld (1988), for example, argued that teaching methods that focus on standard textbook questions encourage the development of *procedural* knowledge that is of limited use in nonschool situations. These and similar arguments have contributed toward the growing support for teaching open or process-based forms of mathematics. Supporters of process-based work argue that if students are given open-ended, practical, and investigative work that requires them to make their own decisions, plan their own routes through tasks, choose methods, and apply their mathematical knowledge, the students will benefit in a number of ways. In this study I was particularly interested to discover whether different forms of teaching would create different forms of learning that might then cause students to interact differently with the demands of new and unusual situations.

RESEARCH METHODS

To contrast content- and process-based mathematical environments, I conducted 3-year case studies in two schools. The case-study approach allowed me to monitor the relationships between the students' day-to-day experiences in classrooms and their developing understanding of mathematics. As part of the case studies, I performed a longitudinal cohort analysis of a "year group" of students in each school while they moved from Year 9 (age 13) to Year 11 (age 16). To understand the students' experiences of mathematics, I observed approximately 100 lessons in

This chapter is adapted from Boaler, J. (1998). Open and closed mathematics: Student experiences and understandings. *Journal for Research in Mathematics Education*, *29*, 41-62.

This research was funded by the Economic and Social Research Council, U.K.

each school. I interviewed approximately 20 students and 4 teachers each year; I analyzed comments elicited from students and teachers about classroom events; I gave questionnaires to all students in my case-study year groups ($n \approx 300$); and I collected an assortment of background documentation. These methods, particularly the lesson observations and student interviews, enabled me to develop a comprehensive understanding of the students' experiences and to begin to view the world of school mathematics from the students' perspectives. In addition, I gave the students various assessments during the 3-year period. I designed some of these myself, but I also analyzed school and external examinations, such as the national mathematics examination (General Certificate of Secondary Education [GCSE]), taken by almost all 16-year-olds in England.

THE TWO SCHOOLS

The two schools in the research study were chosen because their teaching methods were very different but their student bodies were very similar. I first chose a school that used a process-based-mathematics approach, a practice that is rare in the United Kingdom; I refer to this school as Phoenix Park. I then selected a school that used a content-based-mathematics approach and that had a student body that was almost identical to that of Phoenix Park. I refer to this school as Amber Hill. The two schools both lie in the heart of mainly White, working-class communities located on the outskirts of large cities.

About 200 students were in the year group I followed at Amber Hill, and approximately 110 students were in the year group I followed at Phoenix Park. Seventy-five percent of Amber Hill students and 76% of Phoenix Park students were below the national average, based on results of National Foundation for Educational Research (NFER) tests, and no significant differences were found between the ethnic, gender, or social-class composition of the two sets of students. At the beginning of Year 9 (the start of my research), I gave the students a set of seven short questions assessing various aspects of number work. Some of these questions were contextualized; that is, they presented mathematical problems within described situations. No significant differences were noted in the performance of the two sets of students on these tests.

Amber Hill School

Amber Hill is a large, mixed, comprehensive school run by an authoritarian head teacher who tried to improve the school's academic record by encouraging and, when possible, enforcing traditionalism. As a result, the school was unusually ordered and orderly, and visitors walking the corridors would observe quiet and calm classrooms with students sitting in rows or small groups, usually watching the board or working through exercises. The students worked through textbooks in every mathematics lesson in Years 9 to 11, apart from approximately 3 weeks of each of Years 10 and 11, when they were given an open-ended task. The eight teachers of mathematics at Amber Hill were all committed and experienced. On the basis of their entry scores and teachers' beliefs about their abilities, the students were grouped into eight sets: Set 1 comprised the highest ability students. In my lesson observations at Amber Hill school, I was repeatedly impressed by the motivation of the students, who would work through their exercises without complaint or disruption.

Unfortunately, control and order in the mathematics classroom do not, on their own, ensure effective learning. My lesson observations, interviews, and questionnaires all showed that many students found mathematics lessons in Years 9, 10, and 11 extremely boring and tedious. In the lessons I observed, the students often demonstrated a marked degree of disinterest in, and a lack of involvement with, their work. In questionnaires and interviews a large proportion of the students across all eight mathematics sets claimed that mathematics lessons were too similar and monotonous, even though the teachers of these groups were quite dissimilar and varied in popularity and experience. The aspect that united the teachers was their common method of teaching: a 15–20 minute demonstration of mathematical methods followed by the students' working through questions in their textbooks, either alone or with their seating partners.

As a result of approximately 100 lesson observations, I classified a variety of behaviors that seemed to characterize the Amber Hill students' approaches to mathematics. One of these, I termed *rule following* (Boaler, 1996). Many of the Amber Hill students held a view that mathematics was all about memorizing a vast number of rules, formulas, and equations, and this view appeared to influence their mathematical behavior. As one student explained, "In maths there's a certain formula to get to, say from A to B, and there's no other way to get to it, or maybe there is, but you've got to remember the formula; you've got to remember it" (Neil, Year 11, Set 7). Another comment typical of these students is "In maths you have to remember; in other subjects you can think about it" (Louise, Year 11, Set 1).

A related aspect of the Amber Hill students' behavior I refer to as *cue-based behavior*. Often during lesson observations, I witnessed students' basing their mathematical thinking on what they thought was expected of them rather than on the mathematics within a question. I was aware that the Amber Hill students used nonmathematical cues as indicators of the teacher's or the textbook's intentions. The teachers believed that the students experienced difficulty because of the closed nature of their primary-school experiences (ages 5 to 11) rather than because of their secondary school teaching.

Phoenix Park School

Phoenix Park school was different from Amber Hill in many respects; most differences derived from the school's commitment to progressive education. The students at the school were encouraged to take responsibility for their own actions and to be independent thinkers. Few school rules existed, and lessons had a relaxed atmosphere. In mathematics lessons at Phoenix Park, the students worked on open-ended projects in mixed-ability groups at all times until January of their final year, when they stopped their project work and started to practice examination techniques. At the beginning of their projects, the students were given a few different starting points among which they could choose, for example, "The volume of a shape is 216, what can it be?" or "What is the maximum-sized fence that can be built from 36 gates?" The students were then encouraged to develop their own ideas, formulate and extend problems, and use their mathematics. The approach was based upon the philosophy that students should encounter a need to use mathematics in situations that were realistic and meaningful to them. If a student or a group of students needed to use some mathematics that they did not know about, the teacher would teach it to them. Each project lasted 2 to 3 weeks, and at the end of the projects the students were required to turn in descriptions of their work and their mathematical activities. Prior to joining Phoenix Park, the students all had attended schools in which they had experienced the same mathematical approach as the Amber Hill students.

A number of people have commented to me that Phoenix Park's approach must be heavily dependent upon highly skilled and very rare teachers, but I am not convinced that it is. The head of department who devised the approach left at the beginning of my research, and the new head of department was ambivalent toward the approach. A second teacher preferred textbooks but tried to fit in. A third teacher believed in the approach but had difficulty controlling his classes and getting them to do the work. The fourth teacher was newly qualified at the start of the research. The four teachers that made up the mathematics department were all committed and hard working, but I did not regard them as exceptional.

In the 80 mathematics lessons I observed at Phoenix Park, very little control or order was imposed, and, in contrast with Amber Hill lessons, the lessons had no apparent structure. Because they were expected to be responsible for their own learning, the students could, if they wished, work unsupervised. In many of my lesson observations, I was surprised by the numbers of students doing no work or choosing to work only for tiny segments of lessons. In Year 10 the students were given questionnaires in which they were asked to write a sentence describing their mathematics lessons. The three most popular descriptions from Phoenix Park students ($n = 75$) were "noisy" (23%), "a good atmosphere" (17%), and "interesting" (15%). These descriptions contrasted with the three most popular responses from Amber Hill students ($n = 163$), which were "difficult" (40%), something related to their teacher (36%), and "boring" (28%). When I interviewed the students at Phoenix Park and asked them to describe their lessons, the single factor with the highest profile was the degree of choice they were given. The students also talked about the relaxed atmosphere at Phoenix Park, the emphasis on understanding, and the need to explain methods.

During lesson observations at both schools, I frequently asked the students to tell me what they were doing. At Amber Hill, most students would tell me the textbook chapter title, and, if I inquired further, the exercise number. I generally had difficulty obtaining any further information. At Phoenix Park, the students would describe the problem they were trying to solve, what they had discovered so far, and what they were going to try next. In lessons at Phoenix Park, the students discussed with one another the meaning of their work and negotiated possible mathematical directions. In response to the questionnaire item concerning remembering or thinking, 35% of Phoenix Park students (compared to 64% of Amber Hill students) prioritized *remembering*. At its best, the Phoenix Park approach seemed to develop students' desire and ability to think about mathematics in ways that the Amber Hill approach did not.

However, not all feedback from Phoenix Park was positive. About a fifth of the students in each group liked neither the openness of the approach nor the freedom they were given. They did not work when left to their own devices, and they said that they preferred working from textbooks. Most of these students were boys, and they were often disruptive, not only in mathematics classes but throughout the school. Some of the girls described the difficulties these students experienced: "Well, I don't think they were stupid or anything. They just didn't want to do the work; they didn't want to find things out for themselves. They would have preferred it from the book—they needed to know straight away sort of thing" (Anna, Phoenix Park, Year 11). "They just couldn't be bothered, really, to find anything out" (Hilary, Phoenix Park, Year 11). Certainly a major disadvantage of the school's approach seemed to relate to the small proportion of students who simply did not like it and chose to opt out of it.

The students' enjoyment of their mathematics teaching seemed to be quite different at the two schools. At Amber Hill, a strong consensus was found about the shortcomings of the school's approach, but students seemed divided about the Phoenix Park approach; most liked it, but some hated it. The consensus at Amber Hill related primarily to the monotony of doing textbook work and the lack of freedom or choice the students were given. In Year 9, 263 students completed a questionnaire in which they described what they disliked about mathematics at school; 44% of the Amber Hill students criticized the mathematics approach, and 64% of these students criticized the textbook system. Other common criticisms included the constant need to rush through work, the tendency for books to go "on and on," the lack of freedom to work at one's own level, and the lack of choice about topics or order of topics. At Phoenix Park,

only 14% of the students criticized the school's mathematics approach; the most common response, from 23% of students, was to list nothing they disliked about mathematics at school. One of the most obvious differences when students were invited to give their own opinions about mathematics lessons was that the Phoenix Park students chose to comment on the interest level of their lessons and their enjoyment of open-ended work, whereas the Amber Hill students were more concerned about their lack of understanding and their dislike of textbooks.

STUDENT ASSESSMENTS

To gain insight into the different ways in which the students used the mathematics they learned in school in real-world situations, I decided to focus on applied activities situated within school. If the students were unable to make use of or sense of their school mathematics in such tasks within school, I expected that they would not make use of this mathematics in the real world when they encountered tasks with even greater complexity of mathematical and nonmathematical variables.

The Architectural Activity

In the summer of their Year 9, approximately half the students in the top four sets at Amber Hill ($n = 53$) and four of the mixed-ability groups at Phoenix Park ($n = 51$) were asked to consider a model and a plan of a proposed house and to solve two problems related to Local Authority design rules. The students were given a scale plan that showed different cross sections of a house and a scale model of the same house. To solve the problems, students needed to find information from different sources, choose their own methods, plan routes though the task, combine different areas of mathematical content, and communicate information. The Amber Hill students, taken from the top groups, had scored significantly higher than the Phoenix Park students on their mathematics entry tests. However my main aim was not to compare the overall performance of the students in the two schools but instead to compare each individual's performance on the applied activity with his or her performance on a short written test. Approximately 2 weeks prior to completing the architectural task, the students were assessed (on a paper-and-pencil test) on all the mathematical content they needed to use in the activity.

The architectural activity comprised two main sections. In the first section the students were asked to decide whether the proposed house satisfied a council rule about proportion. The rule stated that the volume of the roof of a house must not exceed 70% of the volume of the main body of the house. The students therefore needed to find the volume of the roof and that of the house and to find the roof volume as a percentage of the house volume. To do so, the students could use either the scale plan or the model. The second council rule stated that roofs must not have an angle of less than 70°. The students therefore had to estimate the angle at the top of the roof (which was actually 45°) from either the plan or the model, a shorter and potentially easier task than the first. Grades for the two tasks were awarded as follows: A grade of 1 was given if the answer was correct or nearly correct, with one or two small errors; a grade of 2 was awarded if most or all of the answer was incorrect or if the problem was partially attempted. All students made some attempt at the problems.

Three questions in the pencil-and-paper test assessed the mathematics involved in the proportion problem. In these decontextualized questions, the students were asked to find the volume of a cuboid, the volume of a triangular prism (similar to the roof), and a percentage. The students were given a test grade of 1 if they answered all questions correctly and a grade of 2 if they got one or more wrong. The students' results for the two problems are shown with their pencil-and-paper-test results in Table 16.1.

For the roof-volume problem, Table 16.1 shows that at Amber Hill, 29 students (55%), compared with 38 (75%) of the Phoenix Park students, attained a grade of 1 in the activity despite the fact that the Amber Hill students were taken from the top half of the school's ability range. The table also shows that at Amber Hill 15 students (28%) could do the mathematics when it was assessed in the test but could not use it in the activity; 8 such students (16%) were found at Phoenix Park. In addition, 15 students (29%) at Phoenix Park, compared with 6 students (11%) at Amber Hill, attained a 1 on the activity despite getting one or more of the relevant test questions wrong.

In the test on angle, the students were given a 45° angle (the same angle as the roof in the activity) and were asked whether it was 20°, 45°, 90°, or 120°. Fifty Amber Hill students estimated this angle correctly in the test, but only 31 of these students estimated the 45° angle correctly in the applied activity. At Phoenix Park, 40 of 48 students who recognized the angle in the test solved the angle problem. Paradoxically, the least successful students at Amber Hill were in Set 1, the highest group. Unfortunately, the sight of the word *angle* seemed to prompt many of the Amber Hill Set 1 students to think that trigonometry was required, even though this approach was inappropriate in the context of the activity. The students seemed to take the word *angle* as a cue to the method to use.

In their Year 10, 100 students in each school completed another applied task and a short written

Table 16.1
Volume and Angle Results From Activities and Tests

		Amber Hill (n = 53)					Phoenix Park (n = 51)		
						Volume			
		Activity grade					Activity grade		
		1	2				1	2	
Test grade	1	23	15	38	Test grade	1	23	8	31
	2	6	9	15		2	15	5	20
		29	24	53			38	13	51
						Angle			
		Activity grade					Activity grade		
		1	2				1	2	
Test grade	1	31	19	50	Test grade	1	40	8	48
	2	3	0	3		2	2	1	3
		34	19	53			42	9	51

test; the same pattern of results emerged, but these results were more marked. The Phoenix Park students gained significantly higher grades in all aspects of the applied task, and their performances on test and applied situations were very similar. In applied settings, the Amber Hill students experienced difficulty using the mathematics that they could use in a test; this difficulty again related to their choices of methods.

Short, Closed Questions

The fact that the Phoenix Park students gained higher grades in applied, realistic situations may not be considered surprising, given the school's project-based approach. However, in traditional, closed questions, the Amber Hill students did not perform any better than the students at Phoenix Park. In a set of seven short, written tests of numeracy that I devised, no significant differences were found in the results of the two schools, either at the beginning of Year 9 (the beginning of their new approaches) or at the end of Year 10 (2 years later). The General Certificate of Secondary Education (GCSE) examination is made up of short, fairly traditional, closed questions that assess content knowledge, apart from a few questions that are more applied. Students take these examinations at the end of Year 11. Entry into advanced courses of study, as well as into many professional jobs, generally depends on one's gaining a grade of A, B, or C on this examination. A pass at GCSE is any grade from A to G. In their GCSE examinations, 11% of the Amber Hill cohort attained an A–C grade, and 71% passed the examination. At Phoenix Park, 11% of the cohort attained an A–C grade, and 88% passed the examination. Significantly more of the Phoenix Park students than Amber Hill students attained an A–G pass despite the fact that the GCSE examination was markedly different from anything the students were accustomed to at Phoenix Park.

DISCUSSION AND CONCLUSION

The relative underachievement of the Amber Hill students in formal test situations may be considered surprising, both because the students worked hard in mathematics lessons and because the school's mathematical approach was extremely examination oriented. However, after many hours of observing and interviewing the students, I was not surprised by the relative performances of the two groups of students. In the examination, the Amber Hill students encountered difficulties because they found that the questions did not require merely a simplistic rehearsal of a rule or a procedure but instead required students to understand what the question was asking and which procedure was appropriate. The questions further required the students to apply to new and different situations the methods they had learned. In interviews following their "mock" GCSE examinations (in which students are given a previous year's GCSE examination under testing conditions), Amber Hill students were clear about the reasons for their lack of success. In their mathematics lessons, the students were never left to decide which method they should use; instead, they would ask the teacher or try to read cues from the questions or from the contexts in which the questions were presented. In the examination, the students tried to find similar cues, but they were generally unable to do so:

T: You can get a trigger, when she says, like, *simultaneous equations* and *graphs, graphically*. When they say, like—and you know, it pushes that trigger, tells you what to do.

JB: What happens in the exam when you haven't got that?

T: You panic. (Trevor, Year 11, Set 3)

The students' responses to their examinations indicated that their textbook learning had encouraged them to develop an inert, procedural (Schoenfeld, 1985) knowledge that was of limited use to them. They did not try to interpret what to do, because they believed that, in mathematics, "you have to remember." Unfortunately, their examination did not provide cues that would help them access their memories.

In contrast, the students at Phoenix Park were prepared to try any question they met, and their success in their examinations was enhanced by their desire and abilities to think about unfamiliar situations and determine what was required:

T: I think it allows—when you first come to the school and you do your projects, and it allows you to think more for yourself than when you were in middle school and you worked from the board or from books.

JB: And is that good for you, do you think?

T: Yes. It helped with the exams where we had to . . . had to think for ourselves there and work things out. (Tina, Year 11)

Indeed, this perceiving and interpreting of situations seemed to characterize the real differences in the learning of the students at the two schools. When the students were presented with the angle problem in the architectural task, many of the Amber Hill students were unsuccessful, not because they were incapable of estimating an angle but because they could not interpret the situation correctly. In examinations they similarly could not interpret what the questions were asking. The Phoenix Park students were not as well versed in mathematical procedures, but they were able to interpret and develop meaning in the situations encountered:

JB: Did you feel in your exam that there were things you hadn't done before?

A: Well sometimes, I suppose they put it in a way which throws you. But if there's stuff I actually haven't done before, I'll try and make as much sense of it as I can and try and understand it as best as I can. (Arran, Year 11)

The students at Phoenix Park seemed to have developed a predisposition to think about and use mathematics in novel situations, and this tendency seemed to rest upon two important principles. First, the students believed that mathematics involves active and flexible thought. Second, the students had developed the ability to adapt and change methods to fit new situations. This flexibility in approach, combined with the students' beliefs about the adaptable nature of mathematics and the need for reasoned thought, appeared to enhance the students' examination performance. The proportion of Phoenix Park students who passed the GCSE examination was higher than the national average, despite the initial attainment of the cohort and despite the fact that the students had not encountered all areas of mathematics that were assessed in their examination. In previous years, Phoenix Park had entered its students for a new form of examination that rewarded problem solving as well as procedural knowledge. The school achieved greater success on this new form of examination.[1] Unfortunately Conservative politicians in the United Kingdom caused the new examination, which held the potential for important improvements in mathematics education, to be abolished.

Another major difference in the learning of the students at the two schools related to their reported use of mathematics in real-world situations. The students at Amber Hill all spoke about their inability to make use of school-learned methods in real situations because they could not see connections between what they had done in the classroom and the demands of their lives outside the classroom (Boaler, 1996).

JB: When you use maths outside of school, does it feel like when you do maths in school or does it feel . . . ?

K: No, it's different.

S: No way; it's *totally* different. (Keith and Simon, Year 11, Set 6)

The students at Phoenix Park did not see a real difference between their school mathematics and the mathematics that they needed outside of school.

JB: Do you think in the future, if you need to use maths in something, do you think you will be able to use what you're learning now or do you think you will just make up your own methods?

G: No, I think I'll remember. When I'm out of school now, I can connect back to what I done [*sic*] in class so I know what I'm doing. (Gavin, Year 10)

The students at Phoenix Park had been enculturated into a system of working and thinking that appeared to be advantageous to them in new and unusual settings.

In this study I found many indications that the traditional back-to-basics mathematics approach of Amber Hill was ineffective in preparing students for the demands of the real world and was no more effective than a process-based approach for prepar-

---

[1] When Phoenix Park first adopted a process-based approach, it was involved in a small-scale pilot study of a new GCSE examination that assessed process as well as content. In 1994 the School Curriculum and Assessment Authority (SCAA) withdrew this examination, and the school was forced to enter students for a traditional, content-based examination. The proportion of students at the school attaining Grades A–C and A–G dropped from 32% and 97%, respectively, in 1993 to 12% and 84%, respectively, in 1994. The school has now reintroduced textbook work in an attempt to raise examination performance.

ing students for traditional assessments of content knowledge. I noted problems with the Phoenix Park approach, including the fact that some students spent much of their time not working. Despite these difficulties, the Phoenix Park students were able to achieve more in test and applied situations than the Amber Hill students; they also developed more positive views about the nature of mathematics (views not described here because of space limitations). One might dismiss these results or attribute them to some other factor, such as the quality of the teachers at Phoenix Park, but a value of this type of study is the flexibility it allows researchers, using the data that are most appropriate, to investigate the influence of various factors. After spending hundreds of hours in the classrooms at the two schools, after hearing the students' own accounts of their learning, after analyzing more than 200 questionnaire responses for each year, and after considering the results of traditional and applied assessments, I have been able to isolate factors that have and factors that have not been influential in the students' development of understanding. One important conclusion that I believe I am able to draw from this analysis is that using a traditional, textbook approach that emphasizes computation, rules, and procedures—at the expense of depth of understanding—is disadvantageous to students, primarily because it encourages learning that is inflexible, school-bound, and of limited use.

REFERENCES

Boaler, J. (1996). Learning to lose in the mathematics classroom: A critique of traditional schooling practices. *Qualitative Studies in Education, 9*(1), 17–34.

Lave, J. (1988). *Cognition in practice: Mind, mathematics, and culture in everyday life*. Cambridge, MA: Cambridge University Press.

Nunes, T., Schliemann, A. D., & Carraher, D. W. (1993). *Street mathematics and school mathematics*. New York: Cambridge University Press.

Schoenfeld, A. H. (1985). *Mathematical problem solving*. Orlando, FL: Academic Press.

Schoenfeld, A. H. (1988). When good teaching leads to bad results: The disasters of "well-taught" mathematics courses. *Educational Psychologist, 23*, 145–166.

DEVELOPING CHILDREN'S UNDERSTANDING OF THE RATIONAL NUMBERS: A NEW MODEL AND AN EXPERIMENTAL CURRICULUM

Joan Moss and Robbie Case, University of Toronto

Abstract. A new curriculum to introduce rational numbers was devised, using developmental theory as a guide. The first topic in the curriculum was percent in a linear-measurement context, in which halving as a computational strategy was emphasized. Two-place decimals were introduced next, followed by 3- and 1-place decimals. Fractional notation was introduced last, as an alternative form for representing decimals. After instruction, 4th-grade students in the treatment group showed deeper understanding of rational numbers than those in the control treatment. No differences between the 2 groups were found in conventional computation.

THE domain of rational numbers has traditionally been difficult for middle school students to master. Although most students eventually learn the specific procedures that they are taught, their general conceptual knowledge often remains remarkably deficient. In the domain of fractions, for example, the majority of Grade 9 students, when asked to estimate the sum of 11/12 + 7/8, chose 19 or 20 as the answer in a multiple-choice format

The second author, Robbie Case, is now deceased.

This article is adapted from Moss, J., & Case, R. (1999). Developing children's understanding of the rational numbers: A new model and an experimental curriculum. *Journal for Research in Mathematics Education, 30,* 122–147.

The research reported in this article was supported by grants from the Social Sciences and Humanities Research Council of Canada and the J. S. McDonnell Foundation.

(Carpenter, Corbitt, Kepner, Lindquist, & Reys, 1980). In the domain of decimals, most middle school graduates asserted that "large" numbers, such as 0.1814, are bigger than "small" numbers, such as .3 or .385 (Hiebert & Wearne, 1986). Percents appear to be no easier: When asked to compute 65% of 160, the majority of high school students either failed to give any answer at all or gave answers that were off by more than an order of magnitude (Moss, 1997). These errors reveal a profound lack of conceptual understanding, extending across all three symbolic representations of rational numbers and calling into serious question existing methods of teaching these representations.

Our general goal in this project was to create curricula that help children develop better overall conceptions of the rational number system as a whole and the way its various components fit together—not just better understandings of one or another of these components in isolation. We based our study on information from developmental psychology (Case, 1985) and from our own intuitions, which are based on our backgrounds in psychology. Our first intuition was that what matters is that the general sequence of instruction remain closely in tune with children's original understandings. Another intuition was that the teaching of one form of representation in some depth is preferable to the superficial teaching of several forms at once. And finally, a third intuition that underpinned our curriculum was that students' informal understandings of proportionality would serve as a useful starting place for their learning of rational number concepts and would help them to gain the overall understanding of this number system that was our goal.

In keeping with these intuitions, we used the percent representation as an introduction to rational numbers—our decision to use this representation will be elaborated in a following section—and presented children with a sequence of tasks that we thought would maximize connections with their original, intuitive understandings of percent. We then gradually introduced decimal and fraction representations and their relationships to percents. Our general procedure was as follows:

1. The visual prop that we selected at the outset was a beaker of water. People have little trouble either in seeing such objects in global, proportional terms (i.e., as being full, nearly full, about half full, nearly empty, or empty) or in selecting proportional rather than absolute matches for them. In these first exercises children were asked to think about relative heights of water in terms of fullness and percent.

2. While the lessons were progressing, we encouraged the children to coordinate their intuitive understandings of percents in this context with their strategies for manipulating the numbers from 1 to 100. The two strategies that we emphasized were *numerical halving* (100, 50, 25, etc.) and *composition* (e.g., 100 = 75 + 25).

3. Once the children understood how percentage values could be computed numerically, we introduced them to two-place decimals. We did so in a measurement context by explaining that a two-place decimal number indicates the percentage of the way between two adjacent whole-number distances that an intermediate point lies (e.g., 5.25 is a distance that is 25% of the way between 5 and 6).

4. Finally, we presented the children with exercises in which fractions, decimals, and percents were to be used interchangeably.

We emphasize that in introducing children to the rational number domain in the order described—percents, decimals, then fractions—we were aware that we were reversing the normal order for introducing these representations. Our decision to do so was based on the following considerations:

1. By the age of 10 or 11, children have well-developed qualitative intuitions regarding proportions; they also have well-developed intuitions about the numbers from 1 to 100. By beginning with percents, we allowed them to bring these two sets of intuitions together in a natural fashion.

2. By beginning with percents, we were able to let the children use a form of representation with which they were already familiar. The children appeared already to know a good deal about percents from their everyday experiences. Before we began the instruction, we asked the children if they had ever heard percentage terminology used in their homes or daily lives. They were able not only to volunteer a number of different contexts in which percents appeared (their siblings' school marks, price reductions in stores having sales, and tax on restaurant bills were the ones most frequently mentioned) but also to indicate a good qualitative understanding of what different numerical values "meant," for example, that 100% meant "everything," 99% meant "almost everything," 50% meant "exactly half," and 1% meant "almost nothing." Beginning with percents rather than fractions or decimals allowed us to capitalize on the children's preexisting knowledge regarding the meanings of these numbers and the contexts in which they are important.

3. By beginning with percents rather than fractions or decimals, we postponed the difficulty of having to compare or manipulate ratios with different denominators, thus allowing the children to concentrate on developing their own procedures for comparison and calculation instead of requiring them to struggle to master a complex set of algorithms or procedures that might seem foreign to them.

4. Every percentage value has a corresponding fractional or decimal equivalent that is easy to determine. The converse, however, is not true. Simple fractions, such as 1/3 and 1/7, have no easily calculated equivalent as percentages or decimals. By beginning with percents, we allowed the children to make their first conversions among the different systems in a direct and intuitive fashion and thus to develop a better general understanding of how the three systems are related.

OUR METHOD FOR UNDERTAKING OUR STUDY

Drawing on the foregoing analysis, we first developed an experimental curriculum for teaching rational numbers. We then conducted a study in which a class of 16 fourth-grade children were taught from our experimental curriculum and another class of 15 fourth graders received more traditional instruction on rational numbers. The two classes of children were very much alike in terms of mathematical ability as measured by standardized tests and were drawn from the same neighborhood. They were taught from the same textbook except for the lessons on rational numbers. The experimental group received twenty 40-minute instructional sessions spread over a 5-month period and taught by one of the researchers (Joan Moss). The control group received twenty-five 40-minute lessons spread across a slightly shorter time interval, again during the regularly allotted period for mathematics. Both teachers attempted to limit their work with rational numbers to the periods set aside for that purpose.

To compare the two groups, we designed a detailed interview (The Rational Number Test) to assess the children's conceptual understanding of fractions, decimals, and percents and of the relationships among them. We interviewed each child before and after instruction on rational numbers.

The experimental curriculum

The experimental curriculum began with exercises in which the children used percentage terminology to describe the fullness of different containers of water ("Approximately what percentage of this beaker do you think is full?") or to guess the level of liquid in a container filled to a particular percentage value ("Where will the liquid come to in this beaker when it is 25% full?"). The children's natural tendency when confronted with the fullness problems was to use a *halving* strategy, that is, to determine where a line representing 50% would go on the cylinder or rectangle, then 25%, then 12.5%, and so on. This strategy was encouraged whenever possible. In the course of the instruction, we noticed that in figuring out the location for the 75% mark on a container, children spontaneously decomposed this value into 50% and 25%, then figured out each of these quantities via *halving*, then added the resultant amounts. This strategy, too, was encouraged.

When we introduced numerical problems for which precise calculations had to be made (problems for which children had to compute, for example, the amount of liquid that would be required to fill a 900-ml bottle 75% full), the children spontaneously used these same strategies. That is to say, they spontaneously began by calculating 50% of 900 ml (450 ml) and 50% of 450 ml (225 ml), then added these two values. To facilitate this process, we initially presented problems that could be solved precisely using this general strategy. Problems that

could be solved by calculating 10% or some multiple of 10% were introduced next, in the contexts of menus, tips, and tax.

In attempting to solve such problems, students began with a strategy that used halving to calculate a precise value, 12.5%, and then estimated 10%, as "a slightly smaller number." Subsequently, to compute more precisely, several students began to draw on their knowledge about money. For example, one student said that because one dollar has 100 cents, then 10% of one dollar is 10¢ and 10% of 200 is 20. When such problems were discussed, the conventional 10% strategy (i.e., divide by 10) gradually became established as an alternative to the halving strategy, for certain kinds of questions.

Once the children were comfortable in solving problems of both sorts, we introduced them to two-place decimals, using percents as the entrée. Large laminated number lines, with each number set 1 meter from the previous one, were placed on the classroom floor. Students were asked to walk some percentage of the distance between two adjacent numbers; it was explained that the total distance they had traveled could be represented with a two-place decimal number in which the whole number represented the number of meters walked and the decimal number represented the percentage of the distance to the next meter mark (e.g., "When you pass the 2-m mark and walk 75% of the way to the 3-m mark, the point you reach can be written as 2.75 m").

Although the foregoing context was arbitrary, the children seemed to understand it immediately and were eager to apply their preexisting knowledge of percents to decimal numbers. The exploration of decimals continued with the use of LCD stopwatches with screens that displayed seconds and hundredths of seconds. These values, which appeared on the screen as decimal numbers, were interpreted as temporal analogs of distance, that is, as numbers that indicated the percentage of time that had passed between any two whole-second values. A number of exercises and games were presented to build up the children's intuitive sense of small time intervals and their facility in representing these intervals in this format (e.g., "Try to start and stop the watch as quickly as possible, three times in a row. What was your shortest time? How does this compare to the shortest times in your group?"). Because the shortest times were often in the range of 9 to 15 centiseconds, these problems presented the children with frequent opportunities for meaningful comparison of numbers (e.g., 0.09 and 0.15) that the literature shows they tend to misrepresent. The exercises were extended so that these time intervals were represented as percents, decimals, and fractions of a second.

A special word must be said about fractions. Fraction terminology was used throughout the program, but only in relation to percents and decimals. At the beginning, all the children naturally used the term *one half* interchangeably with 50%, and most knew that 25% could be expressed as *one quarter* or as *one fourth*. We also told them that the 12 1/2% split was called *one eighth* and showed the children the fraction symbol 1/8. In the final lessons of the experimental instructional sequence, we conducted a lesson in which we made fractions the focus. After that lesson, the students were involved in solving and posing various challenges with mixed representations. These activities included (a) *true or false exercises,* such as "0.375 is equal to 3/8, true or false?"; (b) *stopwatch games* with instructions, for example, "Stop the watch [the LCD stopwatches described above] as close to the sum of (1/2 + 3/4) as possible, and then figure out the decimal value for how close you were"; and (c) *Challenge Addition,* in which students were asked to invent a long, mixed-addition problem, such as 1/4 + 25% + 0.0625 + 1/16, with which to challenge their classmates (or the teacher, an even more popular activity).

The curriculum in the control class

The control class devoted a slightly longer time than the experimental group devoted to the study of rational numbers but followed the program from a widely used Canadian mathematics textbook series, *MathQuest* (Kelly, 1986). The textbook followed the traditional pattern of first introducing fractions, then decimals, and then percents. The rules for addition and subtraction of decimals, as well as for multiplication of one- and two-place decimals, were taught explicitly, with careful attention to the significance of place value. The use of a fraction as an operator and computations involving division of decimals were taught at the end of the sequence.

The assessment-interview tasks

In the interview, the children were presented with 41 items in the pretest interview and 45—the original 41 and 4 new items—in the posttest interview. The test was ordered as follows: 12 items on percents, 13 items on fractions, and finally 16 items that featured decimal questions. The posttest interview was expanded to include 1 more item on percents and 3 more items that featured fractions or decimals. To analyze the data, we assigned all but 4 of the questions to six subcategories: (a) *nonstandard computation* (e.g., Another student told me that 7 is 3/4 of 10—Is it?), (b) *compare and order* (e.g., Which is bigger, 0.20 or 0.089?), (c) *misleading appearance* (e.g., Shade 3/4 of this pie [already portioned into eighths, not fourths]), (d) *word problems* (e.g., A CD is on sale. It has been marked down from $8.00 to $7.20. What is the discount as a percentage of the original price?), (e) *interchangeability of representations* (e.g., What is 1/8 as a decimal?), and (f) *standard computation* (e.g., What is .5 + .38?). In assembling the overall battery, we intentionally included a number of questions that were close in their content to the sort of training that the experimental group received and a number that were closer to the training received by the control group.

THE RESULTS OF OUR STUDY

Both groups showed some improvement, but the improvement of the experimental group was significantly greater. The mean scores moved from 12.36 to 31.12 for the experimental group and from 10.79 to 17.5 for the control group. (The pretest had 41 items; the posttest had the same items but included 4 additional items.)

Nonstandard computation

On the items requiring nonstandard computation, the treatment group achieved a mean score at posttest of 6.34 (out of 9) compared to 2.46 for the control group. This difference was statistically significant. The following two test items illustrate differences in the explanations provided by the two groups:

Experimenter: Another student told me that 7 is 3/4 of 10. Is it?

Experimental S1: No, because of one half of 10 is 5. One half of 5 is 2 and 1/2. So if you add 2 1/2 to 5, that would be 7 1/2. So 7 1/2 is 3/4 of 10, not 7.

Experimental S2: No, because a quarter of 10 is 2 1/2, and 2 1/2 times 3 isn't 7; it is 7 1/2.

Control S1: No . . . 7 is not right because it is an odd number, so 6 would be right.

Control S2: Yes, 7 is 3/4 of 10 because 3 plus 4 equals 7.

Experimenter: What is 65% of 160?

Experimental S1: Fifty percent [of 160] is 80. I figure 10%, which would be 16. Then I divided 16 by 2, which is 8 [5%], then 16 plus 8, um . . . 24. Then I do 80 plus 24, which would be 104.

Experimental S2: Ten percent of 160 is 16; 16 times 6 equals 96. Then I did 5%, and that was 8, so . . . 96 plus 8 equals 104.

Control S1: The answer is 95 because 160 minus 65 equals 95.

Control S2: One hundred and sixty divided by 65 equals 2 remainder 30—Is the answer 2?

A complete list of the items in this category, together with the percentage of students in each group that passed each item, is presented in Table 17.1.

Table 17.1
Percentages of Students Succeeding on Items Requiring Some Form of Nonstandard Computation

	Experimental		Control	
Items	Pretest	Posttest	Pretest	Posttest
How much is 50% of $8.00?	89	100	62	92
What is 25% of 80?	34	93	62	92
15 is 75% of what?	50	88	46	46
Is 7 three quarters of 10?	6	75	0	8
How much is 10% of $.90?	19	88	31	23
What is 1% of $4.00?	19	69	0	15
What is 65% of 160?	0	69	0	0
What is .05 of 20?	6	33	8	0
What is 1/2 of 1/8?[a]	—	63	—	0

[a]New item, administered at posttest only.

Compare-and-order problems

The children in the experimental group also showed significantly greater improvement on this class of item than the control group: The experimental group achieved a mean score of 4.75 (out of 7), contrasted with 3.38 for the control group, on the posttest. The responses to the following two items from this category illustrate the different types of understandings that students in the two groups demonstrated and the strategies they used:

Experimenter: Can you think of a number that lies between decimal 3 and decimal 4?

Experimental S1: Well, point three five is between point three and point four.

Experimental S2: Decimal three zero nine.

Control S1: There is no number between decimal three and decimal four.

Control S2: Point zero three.

A second item from this category, "Draw a picture to show which is greater, 2/3 or 3/4," also demonstrated differences between the two groups in their abilities to compare rational number quantities. The solution strategy used by student S1 in the experimental group is representative of the general approach that the students from this group used.

Experimental S1: [First the student drew two equal-sized rectangular figures.] Here is 2/3 [he partitioned the first rectangle into three equal portions, top to bottom, and shaded two bottom parts]. Now you do the same thing for 3/4 [he similarly partitioned the second into four equal parts and shaded three parts]. So 3/4 is bigger because it comes to a higher level on this one [second rectangle].

By contrast, the students in the control group had great difficulty with this item. In fact the posttest score for the control group for this item dropped slightly from the pretest score (see Table 17.2). The responses of the following students exemplify the ways that six students in that group responded:

Control S1: They [2/3 and 3/4] are both the same size because they both have one piece missing.

Misleading-appearance problems

Piaget (1970) believed that unless one presented children with tasks that included misleading features, one was assessing not their true understanding but merely their abilities to parrot what they had been taught. We found that both groups showed some improvement on such misleading items but that the experimental group showed significantly more improvement, attaining a mean score on the posttest of 5.81 (out of 7) as opposed to 2.46 for the control group. On one item, students were asked to shade 3/4 of a pizza that was partitioned into eight sections. The majority of students in the experimental groups responded like one of the following:

Table 17.2
Percentages of Students Succeeding on Items Requiring Comparison and Ordering of Numbers

Items	Experimental		Control	
	Pretest	Posttest	Pretest	Posttest
Which is fewer, 1/3 or 1/2 of 6 blocks?	44	100	69	92
Is there a number between 0.3 and 0.4? Can you name one?	44	100	31	15
Which is bigger, 0.20 or 0.89?	38	81	46	46
Draw a picture to show which is greater, 2/3 or 3/4.	31	81	46	38
Place the fractions 1/2 and 1/3 on a number line.	25	75	31	54
Which is bigger, tenths, hundredths, or thousandths?	6	56	23	85
What is more, 0.06 of 1/10 or 0.6 of 1/100?	0	44	0	15

Experimental S1: I don't know.... Well, let me see.... This is a half [student shaded four sections], . . . so you would need two more to make 3/4 [shades two more sections].

Experimental S2: There are two slices in a quarter, so you need six [slices] to make three quarters [shades them].

Experimental S3: [Shades six sections] I just keep the quarters and forget about the eighths.

But the control students were more likely to explain this way:

Control S1: It says 3/4, so you need to shade in three parts [shades them].

The percentages of students correctly solving the items in this category are shown in Table 17.3.

Word problems

This category comprised only three test items. As can be seen in Table 17.4, the students in the experimental group were successful than those in the control group with these questions, achieving a mean posttest score of 2.56 (out of 3) as opposed to 1.5 for the control group. The contrast in the justifications provided by the two groups is illustrated below:

Experimenter: A CD is on sale. It has been marked down from $8.00 to $7.20. What is the discount as a percentage [of the original price]?

Experimental S1: I knew it was 80 cents. I did the quick math in my head, and I figured out 80 cents was 10%.

Control S1: Eight take away seven point two zero is 80. So 80%.

Table 17.3
Percentages of Students Correctly Solving Items With Misleading Visual Features

Items	Experimental		Control	
	Pretest	Posttest	Pretest	Posttest
Find 3/4 of a pizza (predivided into 8ths).	75	100	38	54
Here is Mary, on her way to school. What fraction of the distance has Mary travelled from her home to school?[a]	25	93	23	31
What percentage of the distance has she travelled?	25	82	8	38
Construct the number 23.5 with base-10 blocks.[b]	56	82	46	46
Can you tell me what number should go at point B on the number line below?	6	44	0	23
Can you tell me what number should go at point A on the number line below?	6	44	8	15

0   **B**   0.1   **A**   0.2   0.3

Shade 0.3 of a circle (predivided into 5 sections).	6	25	8	23

[a]On this problem the units into which the route is divided are small and close to each other; children tend to get very involved in counting the correct number of units Mary has traveled so far (5), and they neglect to count the total distance. The correct answer is 5/8; the most common incorrect answer is 1/5.

[b]The misleading feature here is that the long rods must be used to represent ones, and the centicube blocks, to represent tenths. Students are more familiar with centicube blocks as representing units.

Table 17.4
Percentages of Students Correctly Solving Word Problems on Pretest and Posttest

	Experimental		Control	
Items	Pretest	Posttest	Pretest	Posttest
A book has 100 pages. When Jon is 90% finished reading the book, how many pages has he read?	100	100	100	100
Joan is 100% taller than her daughter Jessica. Jessica's height is what percentage of Joan's? (Ans: 50%)	50	100	31	62
These CDs are marked down from $8.00 to $7.20. What is the discount as a percentage of the original price?	6	56	0	8

The percentages of children correctly solving the three problems in this category are presented in Table 17.4.

Interchangeability of representations

The experimental group showed significantly more improvement than the control group on items requiring translation among fractions, decimals, and percents. The experimental group's mean score on the posttest was 5.88 (out of 8) compared to 2.72 for the control group. The strategies that students used are illustrated here:

Experimenter: What is 1/8 as a decimal?

Experimental S: Zero point one two five.

Experimenter: How did you get that?

Experimental S: Well, 1/4 is 25%, . . . and 1/8 is half of that, so it is 12 1/2%. . . . So 12 1/2% is decimal one two and one half . . . or decimal one two decimal five. No, I think it is just decimal one two five.

Control S: I think it is decimal eight.

Experimenter: How did you get that?

Control S: Because 1/8 is probably the same as decimal eight.

The percentages of children who were successful on the problems in this category are presented in Table 17.5.

Standard computation

Five items were grouped because they included a standard form of computation. On these items, the posttest performance of the two groups was similar (1.94 out of 6 for the experimental group versus 1.54 for the control group), and the statistical analysis did not reveal any significant difference. Still, because the experimental group had not been taught any formal algorithms, we were interested in a number of differences in the ways the items were approached. The following protocols illustrate the nature of these differences:

Experimenter: What is 3 1/4 minus 2 1/2?

Experimental S: I have to carry it over, but I don't know how to carry it over, but since I'm doing a whole, shouldn't we use a quarter and a whole and then subtract a half? So the answer would be 3/4.

Control S: First I must find the common denominator, which is 4; therefore it would become 3 1/4 minus 2 2/4, which equals 1 0/4. So 1 3/4 minus 1 0/4; 3/4.

A complete list of the standard computation items and the percentage of children who gave

Table 17.5
Percentages of Students Succeeding on Items Requiring Movement Among Different Rational Number Representations

	Experimental		Control	
Items	Pretest	Posttest	Pretest	Posttest
How much is 50% of $8.00?	88	100	62	92
How many is 0.5 of all the blocks?	56	100	77	77
What is 1/8 as a decimal?	6	75	0	0
What is 1/3 as a percent?	19	69	8	23
What is seventy-five thousandths as a decimal?	0	25	8	31
How would you write 6% as a fraction?	0	13	15	15
What is 6% as a decimal?[a]	—	93	—	23
What is 1/8 as a percent?[a]	—	75	—	0
What is thirty-five hundredths as a decimal?[a]	—	25	—	46

[a]New item that was administered at posttest only.

Developing Children's Understanding of the Rational Numbers

Table 17.6
Percentages of Students Successfully Completing Items Involving Standard Computation

Items	Experimental		Control	
	Pretest	Posttest	Pretest	Posttest
What is 1/3 of 15?	63	88	62	69
How much is 0.5 + 0.38?	6	50	8	38
How much is 3.64 – 0.8?	0	44	8	38
How much is 2/3 of 6/7?	0	0	0	0
What is 7 1/6 – 6 1/3?	0	0	0	0
What is 3 1/4 – 2 1/2?	0	38	0	0

correct answers for each item are presented in Table 17.6.

CONCLUDING NOTES

As long as they were asked to perform only standard procedures for manipulating very simple numbers, the children in the control group did reasonably well, both in the training and on the posttest. However, when they were confronted with genuinely novel problems, particularly ones for which a misleading cue had to be overcome or some new procedure had to be generated, they continued to make the classic mistakes that they had made on the pretest and that have been reported in past research. Most of these mistakes involved some sort of confusion of the rational numbers with whole numbers. Thus, the responses from the control group were symptomatic of the problem that was cited at the outset of the present article: Although we educators do succeed in teaching children to manipulate rational numbers with our current instructional methods, we fail to help them develop a deep conceptual understanding of these numbers and overcome the fundamental misconception with which they start out their learning, namely, that rational numbers are just special kinds of whole numbers.

Like the control group, the experimental group showed some improvement on the problem subset that required the application of conventional algorithms. In contrast with the control group, however, they showed a large and statistically significant improvement on the other problem sets. They not only got more answers correct than the control group on these other subsets but also reasoned about the problems in ways that were qualitatively different and that demonstrated a deeper, more proportionally based understanding. Indeed, even when the experimental-group students incorrectly answered

a problem on the posttest, they usually made errors in which they respected the proportional status of the underlying numbers. Once again, this result was in sharp contrast with results from the control-group students, who continued to treat the underlying entities as though they were whole numbers.

The approach that we developed for teaching rational numbers was to begin with percents and decimals rather than fractions, and we focused on benchmark equivalencies among percents, decimals, and fractions throughout the curriculum. Our conjecture is that if these techniques were adapted on a more widespread basis, educators would be better able to capitalize on children's accomplishments by the time they reach middle school, on the one hand, and to lay a more solid foundation for the future, on the other. Additionally, we think that we have shown that the usual sequencing of topics in mathematics is not sacrosanct and deserves to be rethought in some instances.

REFERENCES

Carpenter, T. P., Corbitt, M. K., Kepner, H. S., Jr., Lindquist, M. M., & Reys, R. E. (1980). National assessment: A perspective of students' mastery of basic mathematics skills. In M. M. Lindquist (Ed.), *Selected issues in mathematics education* (pp. 215–227). Chicago: National Society for the Study of Education, and Reston, VA: National Council of Teachers of Mathematics.

Case, R. (1985). *Intellectual development: Birth to adulthood.* Orlando, FL: Academic Press.

Hiebert, J., & Wearne, D. (1986). Procedures over concepts: The acquisition of decimal number knowledge. In J. Hiebert (Ed.), *Conceptual and procedural knowledge: The case of mathematics* (pp. 199–223). Hillsdale, NJ: Erlbaum.

Kelly, B. (Series Ed.). (1986). *Mathquest.* Don Mills, ON: Addison-Wesley.

Moss, J. (1997). *Developing children's rational number sense: A new approach and an experimental program.* Unpublished master's thesis, University of Toronto, Toronto, ON.

Piaget, J. (1970). Piaget's theory. In P. H. Mussen (Ed.), *Carmichael's manual of child psychology* (3rd ed., Vol. 1, pp. 703–732). New York: Wiley.

A COMPARISON OF INTEGER ADDITION AND SUBTRACTION PROBLEMS PRESENTED IN AMERICAN AND CHINESE MATHEMATICS TEXTBOOKS

Yeping Li, University of New Hampshire, Durham

Abstract. Integer addition and subtraction problems in several U.S. and Chinese mathematics textbooks were compared to illuminate the cross-national similarities and differences in expectations related to students' mathematics experiences in the United States and China. The differences found in problems' performance requirements indicate that the U.S. textbooks included more variety in problem requirements than the Chinese textbooks. The results are relevant to documented cross-national differences in U.S. and Chinese students' mathematical performances.

EFFORTS to explore possible contributing factors in cross-national differences in students' mathematical achievement have led to the discovery that curriculum is a key contributing factor (e.g., Schmidt, McKnight, & Raizen, 1997). Researchers have analyzed textbooks to understand their potential effect on students' mathematical achievement in the United States and other countries (e.g., Mayer, Sims, & Tajika, 1995). But these studies focused on content analyses, including content-topic coverage and page space devoted to each topic, rather than on the analysis of problems presented in the textbooks. In this study I compared the problems presented in several selected U.S. and Chinese middle school mathematics textbooks to find similarities and differences in expec-

tations related to the mathematical experiences of students in the two countries.

Previous studies on mathematical problems have shown that a problem's mathematical and contextual features are two important dimensions for analyzing mathematical problems. For example, Stigler, Fuson, Ham, and Kim (1986) classified addition and subtraction word problems in U.S. and Soviet elementary school textbooks by using a coding scheme that was based on the unknown's position in the equation that represented the story (a mathematical feature) and the type of story action (a contextual feature); they showed that these two features can potentially affect the development of students' mathematical proficiency in solving these problems. The differences embedded in a problem's performance requirements, such as type of response elicited (e.g., explanation) and cognitive aspects (e.g., procedural practice), are a third dimension of problem features that influence students' mathematical performance.

In a previous textbook study, Carter, Li, and Ferrucci (1997) examined features of content presentation (i.e., explanations and worked-out examples; the use of relevant illustrations; the use of irrelevant illustrations, and exercises) in sections on addition and subtraction of integers in selected U.S. and Chinese middle school textbooks. That study showed that although U.S. and Chinese textbooks were similar in percentage of page space devoted to the various features of content presentation, they differed qualitatively in their organization of content and use of representations. The problems in these sections were not examined,

This chapter is adapted from Li, Y. (2000). A comparison of problems that follow selected content presentations in American and Chinese mathematics textbooks (brief report). *Journal for Research in Mathematics Education, 31,* 234–241.

however, for their potential effects on students' mathematical performance. Yet textbook problems, through which students are expected to gain mathematical experience, may reveal some otherwise unknown effects of instruction. Therefore, I conducted this study to examine the value of textbook-problem comparison in cross-national studies. Specifically, I analyzed all relevant problems presented immediately following the introduction of the content on addition and subtraction of integers in several textbooks from the United States and China.

METHOD USED IN THIS STUDY

The data source consisted of relevant lessons on integer addition and subtraction from five U.S. textbooks and four Chinese textbooks. All textbooks were intended for use in the seventh grade in the United States or for its equivalent in the Chinese educational system. All five U.S. textbooks were extensively used across the country in various settings and with diverse populations. The Chinese textbooks were widely used in different geographic areas of China and bore the approval of the State Education Commission.

The manner in which textbooks are used in Chinese classrooms is similar to that in U.S. classrooms. The textbook provides a blueprint for content coverage and instructional sequence, so Chinese teachers generally develop their teaching plans on the basis of textbooks. However, individual teachers have autonomy in using different pedagogical strategies and modifying instructional emphases to facilitate students' learning. The solutions of the mathematical problems presented in textbooks are part of the specific requirements for students when they are learning the corresponding content. I analyzed those mathematical problems or problem components for which no accompanying solutions or answers were presented. The student was implicitly or explicitly instructed in the textbooks to complete these problems to gain practice with integer addition or subtraction or to apply integer addition or subtraction in practical situations. The problems appeared within, or immediately following, the selected content sections, and they appeared under such headings as "Check Understanding," "Exercise," or "Application."

To undertake this analysis, I developed a three-dimensional framework for analyzing problems: I looked at a problem's mathematical features, contextual features, and performance requirements. Because I analyzed problems on integer operations, I specified a problem's mathematical features as the number of mathematical procedures required for solution and the problem's contextual features as the type of contextual information (purely mathematical or set in a context) in a problem, respectively. I specified the dimension of performance requirements according to (a) the problem's response type: numerical answer only, numerical expression only, or explanation or solution required; and (b) its cognitive requirements: procedural practice; conceptual understanding; problem solving; or having special requirements, such as problem posing. For example, the mathematical and contextual features of "Add –16 + 19 + 12" were coded as *multiple procedures required* as its mathematical feature and *purely mathematical context* as its contextual feature. The performance requirements were coded as *answer only, procedural practice*. The problem "Rich has $13 in his saving account. He deposits $25. How much money is in his saving account after the deposit?" was coded as *single-procedure required; illustrative context, answer only required; problem solving*.

RESULTS

Because all the Chinese textbooks but none of the U.S. textbooks presented integer operations within the context of rational number operations, the percentage of problems calling for integer addition and subtraction in the common content sections in these two nations' textbooks varied. Almost all problems (99%) in the selected content sections in U.S. textbooks were about integer addition and subtraction, but only 64% of the problems in the corresponding sections in the Chinese textbooks related to integer addition and subtraction. Only the problems calling for integer operations were analyzed in this study. The results are reported as a comparison between the two countries' textbooks.

In general, the differences in the coded results for the first two dimensions, mathematical and contextual features, were smaller than the differences in performance requirements. In both U.S. and Chinese textbooks, most problems analyzed required a single computation procedure (80% in both U.S. and Chinese textbooks) and had purely mathematical contexts (87% in U.S. textbooks and 90% in Chinese textbooks). But I found striking differences between U.S. and Chinese textbooks with respect to problems' performance requirements.

The problem-response types were classified as a numerical answer only (A), a numerical expression only (E), or explanation or solution required (ES). Clearly, the majority of problems from both nations were found to require a numerical answer (see Figure 18.1). However, 19% of the problems in U.S. textbooks but none of those in Chinese textbooks required ES responses. Because, in the U.S. textbooks, the problems that required a mathematical solution also asked for a mathematical (e.g., pictorial) explanation of the solution process, this difference in explanatory requirement (i.e., 19% in U.S. textbooks and 0% in Chinese textbooks) reflects the emphasis given to enhancing communication skills in U.S. *Standards*-based reform documents (e.g., NCTM, 1989).

In the analysis of cognitive requirement, 26% of the U.S. textbook problems and 16% of the Chinese textbook problems were found to require conceptual understanding (CU) for solution. In contrast, 63% of the U.S. textbook problems and 72% of the Chinese textbook problems were classified as requiring procedural practice (PP). This evidence of increased emphasis on conceptual understanding and reduced attention to procedural practice is also in line with the focus of current U.S. mathematics education reform efforts. The problem-solving requirement (PS) was about the same for the two countries but was low for both. Moreover, although only 2% of the U.S. textbook problems had a special requirement (SR), no such problems were found in the Chinese textbooks. The differences found in cognitive requirement and response type indicate that the U.S. textbooks included more variety in problem requirements than the Chinese textbooks.

DISCUSSION

These results indicate that U.S. textbook problems varied in problem requirements and emphasized conceptual understanding more than Chinese textbook problems. Because 36% of the Chinese textbook problems but almost none of the U.S. textbook problems in content sections on integer addition and subtraction called for the operations on fractions or decimals, I determined that Chinese textbooks included more problems with higher level mathematics content than U.S. textbooks and that Chinese students were expected to acquire the ability to add and subtract rational numbers at a quicker pace than U.S. students. Therefore, the results of this study illustrate the value of conducting textbook-problem comparison, complementing textbook-content comparison.

Differences between U.S. and Chinese textbooks shown by my snapshot comparison of problems in a common content topic include the requirement of mathematical explanation and the emphasis on conceptual understanding embedded in U.S. textbook problems. These differences indicate that U.S. textbooks emphasized the improvement of students' mathematical communication and conceptual understanding. Such results are relevant to documented cross-national differences in U.S. and Chinese students' mathematical performances. For example, U.S. students scored dramatically lower than their Chinese counterparts in solving traditional mathematics problems but not in solving open-ended problems (e.g., Cai, 1995), for which students were asked to provide explanations for their solutions. Therefore, the cross-national differences in the mathematical experience called on to solve textbook problems, as illustrated in this study, provide a basis for proposing an "experience hypothesis" to project students' actual mathemat-

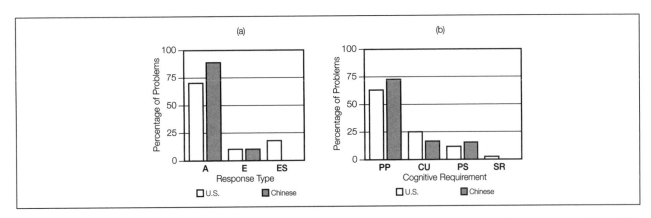

Figure 18.1. Results from coding textbook problems in three dimensions: Performance requirements (response type and cognitive requirement).

ics performances. As an exploratory effort, this study has provided a glimpse into the potential relationship between the expectations for students' experiences and their actual performances in mathematics.

REFERENCES

Cai, J. (1995). *A cognitive analysis of U.S. and Chinese students' mathematical performance on tasks involving computation, simple problem solving, and complex problem solving. Journal for Research in Mathematics Education* Monograph Series, Number 7. Reston, VA: National Council of Teachers of Mathematics.

Carter, J., Li, Y., & Ferrucci, B. (1997). A comparison of how textbooks present integer addition and subtraction in China and the United States. *Mathematics Educator, 2*(2), 197–209.

Mayer, R. E., Sims, V., & Tajika, H. (1995). A comparison of how textbooks teach mathematical problem solving in Japan and the United States. *American Educational Research Journal, 32,* 443–460.

National Council of Teachers of Mathematics. (1989). *Curriculum and evaluation standards for school mathematics*. Reston, VA: Author.

Schmidt, W. H., McKnight, C. C., & Raizen, S. A. (1997). *A splintered vision: An investigation of U.S. science and mathematics education*. Dordrecht, The Netherlands: Kluwer.

Stigler, J. W., Fuson, K. C., Ham, M., & Kim, M. S. (1986). An analysis of addition and subtraction word problems in American and Soviet elementary mathematics textbooks. *Cognition and Instruction, 3,* 153–171.

SUPPORTING LATINO FIRST GRADERS' TEN-STRUCTURED THINKING IN URBAN CLASSROOMS

Karen C. Fuson, Steven T. Smith, and Ana Maria Lo Cicero, Northwestern University

Abstract. Year-long classroom-teaching experiments in 2 predominantly Latino low-socioeconomic-status (SES) urban classrooms (1 English speaking and 1 Spanish speaking) were designed to support 1st-graders' thinking of 2-digit quantities as 10s and 1s. A model of a developmental sequence of conceptual structures for 2-digit numbers (the UDSSI triad model) is presented to describe children's thinking. By the end of the year, most of the children could accurately add and subtract 2-digit numbers that require trading (regrouping) by using drawings or objects and could give answers by using 10s and 1s on various tasks. Their performance was substantially above that reported in other studies for U.S. 1st graders of higher SES and for older U.S. children. Their responses looked more like those of East Asian children than those of U.S. children in other studies.

OUR first purpose in this chapter is to describe two aspects of the research that may be particularly helpful to others: (a) a developmental sequence of conceptual structures for two-digit numbers that guided the instructional-design work

---

This chapter is adapted from Fuson, K. C., Smith, S. T., & Lo Cicero, A. M. (1997). Supporting Latino first graders' ten-structured thinking in urban classrooms. *Journal for Research in Mathematics Education, 28,* 738–766.

The research reported in this chapter was supported in part by the National Center for Research in Mathematical Sciences Education, which is funded by the Office of Educational Research and Improvement (OERI) by the U.S. Department of Education under Grant R117G10002, and in part by the National Science Foundation (NSF) under grant RED935373. The opinions expressed in this chapter are those of the authors and do not necessarily reflect the views of OERI or of NSF.

and (b) the conceptual supports we used to assist children's construction of these conceptual structures. In particular, we describe our method for encouraging children to draw quantities organized by tens; using this method, we addressed major pragmatic and instructional-assessment issues and afforded multiple solution methods. Our second purpose in this chapter is to describe the learning of the children in the two classes as it compared with that of East Asian and other U.S. students. We first describe the developmental sequence of conceptual structures for two-digit numbers.

ANALYSIS OF THE MATHEMATICAL DOMAIN OF MULTIDIGIT NUMBERS

We developed a triad model of two-digit conceptions (shown in Figure 19.1). (The introduction to this section contains additional information that might help you understand various parts of this complex figure.) *Triad* refers to the relationships among quantities, number words, and written numerals. In this model we describe 5 two-digit conceptions.

All children begin with a *unitary conception* of two-digit numbers; this conception is a simple extension from a unitary conception of single-digit numbers (see the top-middle and top-right-hand parts of Figure 19.1). With this conception, the number words and the two digits do not each refer to quantities being counted. Rather, the entire number word (e.g., *fifty-three*) or numeral (53) refers to the whole quantity; that is, the child does not recognize tens and ones.

With time and experience, each number word and each digit do take on meaning as a decade or as the extra ones in the *decade-and-ones* conception.

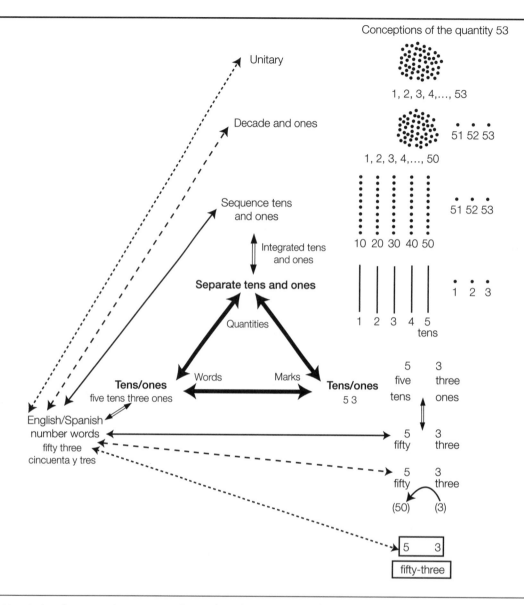

Conceptions of the quantity 53

Figure 19.1. A developmental sequence of conceptual structures for two-digit numbers: The UDSSI Triad Model.

For example, in 53, the 5 means fifty and the 3 means three.

In the *sequence-tens-and-ones conception*, an extension of the decade-and-ones conception, units of ten single units are formed within the decade part of the quantity. These sequence-tens units are counted by tens (e.g., 10, 20, 30, 40, 50); then the ones are counted by continuing from 50 (51, 52, 53).

The *separate-tens-and-ones conception* is built through experiences in which a child comes to think of a two-digit number as comprising two separate kinds of units—units of ten and units of one. Both kinds of units are counted by ones (e.g., "1, 2,

3, 4, 5 tens and 1, 2, 3 ones"). In Figure 19.1, we show these units of ten as a single line to emphasize their (ten)-unitness, but the user of these units understands that each ten is composed of 10 ones and can switch to thinking of 10 ones if that approach becomes useful.

The first two conceptions are developmental, and children move from the first through the second. But children's construction of the sequence-tens-and-ones and separate-tens-and-ones conceptions depends heavily on their learning environments. Students in the same classroom may construct one or the other of these conceptions first: A child who focuses on the words will develop the

156   Lessons Learned From Research

sequence-tens-and-ones conception first; a child who focuses on the written numerals will develop the separate-tens-and-ones conception first.

Children may eventually relate the sequence-tens-and-ones and separate-tens-and-ones conceptions to each other in an *integrated sequence-separate conception* (these connections are shown in Figure 19.1 as the short double arrows). In the integrated conception, children connect *fifty* to 5 tens, and the written numeral 53 can take on either quantity meaning (fifty-three or 5 tens, 3 ones).

We call this model the *UDSSI triad model*, for the names of the five conceptions (unitary, decade, sequence, separate, integrated). We originally thought about each two-digit conceptual structure as a triangle (a triad) of six relations (each two-ended arrow represents two relations). However, we later realized that only the separate-tens-and-ones conception (in bold in Figure 19.1) has direct links between quantities and numerals. Such a direct link can occur only if the quantities of tens and ones are small enough to be subitized (immediately seen as a certain number of units) or are arranged in a pattern. In the other three conceptions, a person must, by counting, relate quantities to written numerals through the number words. Therefore, for these conceptions, the link between quantities and numerals is not drawn in Figure 19.1.

USE OF THE TRIAD MODEL IN OUR STUDY

Our analysis of the structure of Spanish words for two-digit numbers indicates that Figure 19.1 describes the main conceptual structures that Spanish-speaking children construct, albeit with some small advantages and disadvantages compared with English speakers. The relationship between the Spanish decade words (*diez, veinte, treinta, cuarenta, cincuenta, sesenta, setenta, ochenta, noventa*) and the corresponding words for 1 to 9 (*uno, dos, tres, cuatro, cinco, seis, siete, ocho, nueve*) is unclear. This lack of a clear relationship is similar to the rather opaque relations of the English words *twenty* and *two, thirty* and *three,* and *fifty* and *five.* The Spanish words for numbers from 21 onward use a construction (e.g., *cincuenta y tres* for *fifty and three*) that seems to support children's construction of the decade-and-ones conception. Spanish words in the teens are irregular at first and then, at 16, begin to name the ten: *once, doce, trece, catorce, quince, dieciséis, diecisiete, dieciocho, diecinueve.* A less common alternative spelling is conceptually

clearer: *Diez y seis* means "ten and six," whereas *dieciséis* means "tens six." Thus, the conceptual contribution of the words for 16 through 19 may depend heavily on the emphasis given by the teacher to the "ten and six" meaning. Most European systems of number words for two-digit numbers have a decade structure, although with various irregularities. We believe, therefore, that the UDSSI triad model is relevant to these other European languages also.

Our goal in this study was to help children in both English-speaking and Spanish-speaking classrooms construct all the triad conceptual structures. But in what order and how were we to help them? The literature on East Asian children's multidigit learning contributed major elements to the design of the learning activities. Two conceptual structures in Figure 19.1 (the decade-and-ones and the sequence-tens-and-ones structures) come from the decade structure of the English number words. Some countries have number words with no separate list of decade words. For example, in East Asian countries in which the number words are based on ancient Chinese number words, children say 12 as "ten two" and 53 as "five ten three." These children need to construct only the unitary and the separate-tens-and-ones conceptual structures. The construction of the latter triad (shown in bold in Figure 19.1) is supported by the presence in children's lives of ten-structured cultural artifacts and experiences (e.g., the abacus, the metric system), by school instruction on ten-structured methods for adding and subtracting single-digit and multidigit numbers (Fuson & Kwon, 1992), and by the support of parents and teachers in demonstrating ten-structured quantities (Yang & Cobb, 1995). Cross-cultural work on East Asian children's numerical thinking indicates that children build highly effective ten-structured conceptions of numbers; use these very well in their addition and subtraction; and far outperform U.S. children on single-digit and multidigit addition, subtraction, and place-value tasks.

The simplicity of the inner triad indicated to us that for English-speaking children, constructing the separate-tens-and-ones conception might be easier than constructing the decade-and-ones and the sequence-tens-and-ones conceptions. Also, with the separate-tens-and-ones conception, children could participate in classroom activities involving ten-structured quantities before they learned the English (or Spanish) sequence to 100; learning the sequence can take months or even years. To facilitate the learning of the separate-tens-and-ones conception, we decided to use tens-and-ones words to describe quantities. We thought that children's construction of the decade-and-ones and the sequence-tens conceptions might be facilitated by activities that also simultaneously supported the construction of the separate-tens-and-ones conception; such activities include arranging quantities in groups of tens, counting them by tens and ones (a sequence-tens conception), and counting the units of ten and the units of one separately (a separate-tens-and-ones conception). Using both tens-and-ones words (*one ten, two ones*) and ordinary English (*twelve*) or Spanish (*doce*) words would help focus children's attention on both conceptions (i.e., on the separate-tens and on the sequence-tens conceptions).

THE INSTRUCTIONAL PHASE OF THE STUDY

The K–8 school in which the study took place is located in a predominantly Latino neighborhood; 87% of the students qualify for free or reduced-cost lunch. Each grade level from 1 through 5 has a Spanish-speaking and an English-speaking class. Mathematics classes and most other classes are carried out almost entirely in the specified language (English or Spanish). All first graders who had entered the school by mid-December were included in the study sample. The Spanish-speaking first-grade class had 17 children who were in the class from mid-December through June. The English-speaking class had 20 children who were in the class from mid-December through June.

We viewed addition and subtraction work as important settings for children's continued construction of place-value concepts, and thus instruction did not begin with place-value constructions. In solving problems that do not require regrouping, students do not need to use the trading-ten-for-10-ones strategy. Because they do not need to trade in such problems, they often make errors when they later encounter problems that do require regrouping. For example, in subtraction, if instruction begins with solving problems not requiring regrouping, students often make the common top-from-bottom error on problems for which they need to regroup. For these reasons, we began instruction with two-digit problems requiring regrouping.

Finally, our experience indicated that to construct robust ten-structured conceptions, children needed to explore two-digit addition and subtraction using quantities as referents, not using just numerals. Therefore we had to select and design such referents.

We faced various pragmatic constraints concerning our choice of quantity referents to use in the classroom, because we wanted them to be usable in any inner-city classroom. They had to be easy to manage, inexpensive, and require minimal teacher-preparation time. Penny-frames fit these criteria; children fitted pennies in rows of ten, and after a child put a penny into the frame, she or he wrote below that penny the new total number of pennies in the frame. Other referents, such as a hundreds chart, were also occasionally used.

For the initial triad activities, the teacher at first led simultaneous performance by all children; each child did the activity at his or her seat, following or with the teacher. We began with activities designed to help children (a) see objects grouped into tens and (b) relate these tens groupings and the leftover ones to number words and written numerals.

In all classroom activities, the teacher used each of the triad conceptual structures successively in various orders. For example, pennies were placed individually into the penny frame while the teacher counted each one unitarily. When the activity progressed over days, an increasing number of children became able to count with the teacher. As each ten row was filled, the rows of ten would be counted by tens (e.g., 10, 20, 30) and then by ones (1, 2, 3 groups of ten). All the pennies might then be counted again by ones to verify that 30 pennies were found in the three groups of ten. Written numerals would be read as tens-and-ones words, in English words in English-speaking classes and in Spanish words in the Spanish-speaking class: "So 36 is thirty-six pennies, three groups of ten pennies and six loose pennies left over. We write 3 tens here on the left and 6 ones here on the right." Gradually, individuals would learn to do the task alone, with informal help from peers and the teacher.

Base-ten blocks and other object quantities leave no records after class to help teachers assess their students' understanding. We therefore introduced a system of recording quantities as ten-sticks and dots, which children could count by using any of the conceptions. Initially, children made dots in columns of 10 to make a record of objects the class was collecting. They counted by ones while they made these columns of 10 dots. When they had fewer than 10 left, they made a horizontal row of dots, often with a space between the first five dots and the last four dots to facilitate seeing the number of dots. When many children could make such drawings confidently, the columns of 10 dots were connected; the children drew a line through them while the counting by tens or of tens was done. Some children had already spontaneously begun to do this step. Eventually only the vertical stick was drawn to show a ten. These activities occupied part of each class period for about 2 weeks.

DESIGN ISSUES

One research teacher worked in the English-speaking class, and the other research teacher worked in the Spanish-speaking class. When the regular classroom teacher became familiar with an activity, she took over some or all of the teaching. The research-team teachers were in the classrooms 5 days a week from late September through November and then on Mondays, Tuesdays, and Wednesdays for the remainder of the year. On Thursdays and Fridays the classroom teachers continued with their usual approaches to nonproject topics. During the fall, activities on single-digit addition and subtraction to 10 and word problems occupied much of the time.

After considerable discussion with the classroom teachers, the research team decided on, for example, "five tens and three ones" and "cinco dieces y tres unos" for the tens-and-ones words. Beginning in January we tried various activities, including vertical-number-line or number-bar activities. None seemed particularly powerful or interesting to the children. In mid-February, rows of dots were made for each addend; 10 dots were enclosed, if possible, and the answer was then recorded as "one ten and x ones." For a few minutes on many days during the winter and spring, children solved addition and subtraction problems with sums in the teens; they then demonstrated various finger methods they were using to solve the problems. Explicit practice activities focused on the prerequisites for the ten-structured methods (e.g., "How many more to make 10 [with a given number]?" and "10 plus x = ?"). By midyear most children were counting on or using fingers in other ways that could lead to ten-structured methods.

We decided to use the ten-sticks and dots for two-digit addition and subtraction for their advantages as written records and for cost and management reasons. Teachers used this activity periodically from late February through May. In the English-speaking class, the children did triad review activities for 2 days and then spent four classes on two-digit addition. Six classes on two-digit subtraction (in April and May) were followed by six sessions of mixed addition and subtraction problems with and without trading (in May). The time spent on these two-digit activities averaged 30 minutes for each class. The Spanish-speaking children spent about this same amount of time on triad activities, two-digit addition with trading, three-digit triad activities (drawing squares for hundreds), three-digit addition with no trading, and two-digit subtraction with no trading.

Addition and subtraction lessons typically began with instructional conversations in which the teacher elicited children's ideas about methods. Each method was then carried out by the whole class together. The children next worked alone, solving problems by any method they chose. Conversations to facilitate the children's reflections on, and comparisons of, methods were intermixed with periods during which the children worked either alone or spontaneously together. When errors arose, the teacher also identified and discussed them. The children explained why these examples were errors and how to correct them.

Individual interviews with all participating children were carried out in late May and early June. Interview items were selected mainly from other published studies to obtain data on the children in our classes to compare with the data reported in the other studies. The interviews included a large number of items. Therefore, some items were given to all children, and some were given only to a subgroup of the children drawn from across the whole achievement range of the class.

RESULTS

The classroom teachers reported that the children, especially the least advanced, found the regular tens-and-ones words easier to learn than the standard English or Spanish number words. The children did not seem to confuse the two kinds of words; no construction from one kind was carried into the other kind of words.

The children varied in the ways they drew the ten-sticks and dots, enclosed 10 dots, and showed their answers (see the methods in Figure 19.2). When we watched children solve a problem, we saw other differences in the children's methods. For finding the total, some children counted ones first, and others counted tens first. For finding the number of tens, some children counted all, some counted on from the first number of tens, and some used known facts (e.g., "3 tens and 2 tens make 5 tens") to add the tens. Some children integrated the new, enclosed ten (see new groups of 10 created in the addition problems in Figure 19.2) into a sequence count of the total, and others counted it as another ten in their count of the tens.

Many children made rapid progress in subtracting correctly. All children opened a ten by drawing the 10 ones (see two examples at the bottom of Figure 19.2). Correctly taking away the tens was particularly easy: After only two sessions, 17 of 22 children correctly crossed out the correct number of ten-sticks on the quizzes at the beginning of class.

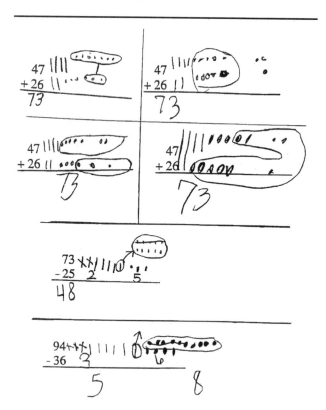

Figure 19.2. Addition and subtraction methods using ten-sticks and dots.

The children's performance on subtraction dropped somewhat when addition and subtraction were mixed. However, by the third class with mixed operations, many children were differentiating their addition and subtraction methods. The children made many more kinds of errors in subtraction than in addition. Nearly every error we could anticipate was made by some child at some time. In addition problems, most errors were in regrouping, and these mistakes were of only a few kinds. In contrast, the subtraction regrouping (opening a ten) seemed to be relatively easy to understand and to carry out with an overall correct approach. However, it was subject to many minor execution errors. Children almost always opened a ten and crossed out the correct number of ones. But miscounting occasionally occurred at each possible counting step, and children sometimes counted only part of the tens or part of the ones.

All children in both classes could do the tasks assessing all six relations for the inner separate-tens-and-ones triad. For the unitary conception, one child in each class could not count to 100 by ones. These children, and an additional child from the English-speaking class, could not count to 100 by tens. All other children did the sequence-tens-and-ones triad tasks correctly. Most children in the Spanish-speaking class did these tasks directly by counting by tens. Many children in the English-speaking class demonstrated the separate-tens-and-ones triad relations and then translated this result to English number words (e.g., counted "1, 2, 3, 4 tens and 1, 2 ones, so that's forty-two").

Many children were able to demonstrate triad relations in the cardinal ten-structured task ("There are 53 first graders at Esperanza School. How many teams of 10 can be made?"). Almost all Spanish-speaking children (94%) and half the English-speaking children answered correctly. Almost all these children (92%) knew the answer rapidly, without counting or drawing.

On various tasks that assess whether children are thinking unitarily or with tens and ones, our first graders from both classes predominantly demonstrated tens-and-ones thinking. Their performance thus looked more like that of East Asian children than that of U.S. children, who predominantly demonstrate unitary or concatenated single-digit conceptions (Miura, Kim, Chang, & Okamoto, 1988). The East Asian children were tested in the first half of the year, whereas ours were tested at

Lessons Learned From Research

the end of the year, so our students are still behind East Asian children in the timing of their use of tens and ones. On Miura and others' task of the cognitive representation of numbers (use blocks to represent a given number), 88% of our students made a ten-structured 42 using 4 tens blocks and 2 units blocks compared with a mean of 89% of children from the People's Republic of China, Japan, and Korea making a ten-structured display (a mean of 10% of the 89% made a noncanonical-ten version that had some tens and more than 9 ones). When asked to make a different block presentation for 42, only 7% of our students made a unitary presentation; a mean of 53% of the East Asian first graders made a unitary presentation. Most of the 74% of our students making a correct second presentation made a noncanonical-ten arrangement in which some ones were arranged in groups of ten.

Sixty-three percent of our first graders immediately said that the *1* in 16 was 10 chips. This percentage is considerably higher than either the 42% of M. Kamii's (1982) 9-year-olds who were correct or the 32% of the second and third graders in our project school before the project began. Our students' 63% correct is about the same as the 60% of C. Kamii's (1985) affluent suburban sixth graders.

The children were given three place-value-understanding tasks that assess children's quantity meanings for each digit in a two-digit number. All tasks and comparison samples are from Miura, Okamoto, Kim, Steere, and Fayol (1993). Our first graders did as well as East Asian children tested in the first half of the year and much better than U.S. first graders from a selective, academically rigorous school with monolingual middle- and upper-middle-class children (that sample and our students were tested at the end of the year).

On the two-digit addition problem (48 + 36) using ten-sticks and dots, 90% of our students' solutions were correct. On the two-digit subtraction problem using ten-sticks and dots, all the children in the English-speaking class correctly opened a ten-stick by drawing 10 enclosed dots and correctly took away the required ten-sticks and dots. Of these, 70% then wrote the correct answer. The Spanish-speaking first graders had not had opportunities in class to solve subtraction problems requiring trading.

Word-problem tasks with base-ten blocks were given to assess whether the ten-structured concep-

tions built by our students would generalize to these unfamiliar tools. On the addition problem, 100% of the Spanish-speaking children were correct, and 90% added using sequence-tens-and-ones counting of the total. Of the English-speaking children, 75% were correct; 84% of these children traded 10 units to make another ten bar, and most of these children counted the tens and ones separately.

On the subtraction problem, all the children showed correct digit correspondence by making 74 with 7 ten-bars and 4 units and trying to take away 3 ten-bars and 8 units. All but two children used a correct strategy for solving this problem and did see the ten-bars both as 1 ten (when making the 74) and as 10 ones (when taking away some or all of the 8 ones from it). Half of these children carried out their strategies correctly, and the rest made some error in executing their strategies.

DISCUSSION

By having the children use ten-stick-and-dot quantity drawing, we were able to support most children's construction of most elements of the conceptual structures shown in Figure 19.1. Furthermore, on a range of unfamiliar tasks, many children showed a robust preference for ten-structured conceptions, performing like children in China, Japan, and Korea rather than like agemates in the United States or like children in higher grades in the United States. Most children were also able to carry out a ten-structured solution to two-digit addition and subtraction problems and to explain their regrouping. First graders in the United States do not ordinarily learn to solve such problems with trades; these problems are usually not included in first-grade textbooks or appear in the final chapter, which many teachers do not reach. Our students' performance was considerably above that reported for U.S. children receiving traditional and reform instruction and was above that reported for Japanese and Taiwanese first graders on some tasks. This superiority is partly because the children's conceptual tool, the ten-sticks and dots, could be drawn on paper and thus could be used on homework or in an assessment whenever pencil and paper were available.

Through some reform approaches to primary school mathematics, many children have been successful in inventing accurate methods of two-digit addition and subtraction and in understanding them. But some children in those projects continue to use unitary methods into second, third, and even fourth grade. Our teacher-orchestrated activities designed to help children construct sequence-tens and separate-tens conceptions and then to use one or the other in two-digit addition and subtraction were quite successful: No first grader used a unitary method. This result shows the benefits in teachers' carrying out such activities with whatever conceptual supports for quantity are used in their classrooms or their introducing ten-sticks-and-dots activities as recordings of any quantities in problems.

In several tasks, the children in the Spanish-speaking class showed a preference for counting by sequence-tens over counting by separate-tens. This preference facilitated their solutions with the unfamiliar media of noncanonical sticks and dots and of the base-ten blocks because they did not have to explicitly make another ten: Counting by tens and then counting the ones took them up over the next decade to get the answer. In contrast, more children in the English-speaking class demonstrated separate-tens-and-ones conceptions in which they had to explicitly make another ten by grouping or adding or had to break a ten. In some new situations, fewer of these children were able to complete this activity accurately. The difference between the two classes in preferred conceptions illustrates how instructional emphases in the uses of a conceptual tool and the uses of different tools can support different conceptual constructions.

The results reported here clearly indicate that all U.S. children can do enormously better than they ordinarily do in primary school mathematics. Furthermore, the widely reported gap in performance and understanding between East Asian children and children in the United States can be narrowed or eliminated, even in poor, inner-city schools. Doing so requires a substantially more ambitious first-grade curriculum and active teaching that supports children's construction of a web of multiunit conceptions in which number words and written number marks (numerals) are related to ten-structured quantities. Drawn quantities, instead of objects, can serve as meaningful ten-structured quantities that support reflection, communication, assistance, and teachers' assessment of children's thinking.

REFERENCES

Fuson, K. C., & Kwon, Y. (1992). Korean children's understanding of multidigit addition and subtraction. *Child Development, 63,* 491–506.

Kamii, C. (1985). *Young children reinvent arithmetic: Implications of Piaget's theory.* New York: Teachers College Press.

Kamii, M. (1982). Children's graphic representation of numerical concepts: A developmental study (Doctoral dissertation, Harvard University, 1982). *Dissertation Abstracts International, 43,* 1478A.

Miura, I. T., Kim, C. C., Chang, C., & Okamoto, Y. (1988). Effects of language characteristics on children's cognitive representation of number: Cross-national comparisons. *Child Development, 59,* 1445–1450.

Miura, I. T., Okamoto, Y., Kim, C. C., Steere, M., & Fayol, M. (1993). First graders' cognitive representation of number and understanding of place value: Cross-national comparisons—France, Japan, Korea, Sweden, and the United States. *Journal of Educational Psychology, 85,* 24–30.

Yang, M. T. L., & Cobb, P. (1995). A cross-cultural investigation into the development of place-value concepts of children in Taiwan and the United States. *Educational Studies in Mathematics, 28,* 1–33.

EFFECTS OF *STANDARDS*-BASED MATHEMATICS EDUCATION: A STUDY OF THE CORE-PLUS MATHEMATICS ALGEBRA AND FUNCTIONS STRAND

Mary Ann Huntley, WestEd/National Center for Improving Science Education
Chris L. Rasmussen, Purdue University—Calumet

Abstract. The vision of *Standards*–based mathematics education was tested in a comparative study of the effects of the Core-Plus Mathematics Program (CPMP) and more conventional curricula on growth of student's understanding, skill, and problem-solving ability in algebra. Results indicate that the CPMP curriculum is more effective than conventional curricula in developing student ability to solve algebraic problems if those problems are presented in realistic contexts and if students may use graphing calculators. Conventional curricula are more effective than CPMP in developing student skills in manipulation of symbolic expressions in algebra if those expressions are presented free of application context and if students do not use graphing calculators.

THE concepts, principles, and techniques of algebra are important tools for describing and reasoning about patterns in all branches of mathematics. Algebra has been at the heart of secondary school mathematics for many years, and high achievement in algebra has long been the hallmark of preparedness for advanced mathematical and scientific studies. Recent recommendations by major mathematics education professional organizations, such as the National Council of Teachers of Mathematics (NCTM, 1989, 1991, 1995, 2000) and the Mathematical Sciences Education Board (National Research Council, 1989), call for fundamental changes in secondary school mathematics curricula, instruction, and assessment. These proposed changes have special implications for the treatment of algebra.

The Core-Plus Mathematics Project (CPMP) is one of several major national efforts funded by the National Science Foundation to construct, implement, and evaluate a high school mathematics program exemplifying principles and practices recommended in these recent proposals for reform. The CPMP curriculum is based on the theme that mathematics is a tool for making sense of the world. Each year of the CPMP curriculum features topics in algebra and functions, geometry and trigonometry, statistics and probability, and discrete mathematics. The seven algebra and functions units, which make up slightly more than one third of the 3-year core curriculum, form one major strand of the Core-Plus Mathematics Program. The authors of the algebra-and-functions strand approach traditional mathematical topics in new ways: emphasizing mathematical modeling; recommending use of graphing calculators to support multiple representations of algebraic ideas; promoting learning through collaborative work on authentic problems; integrating algebra with topics in geometry, statistics, proba-

This chapter is adapted from Huntley, M. A., Rasmussen, C. L., Villarubi, R. S., Sangtong, J., & Fey, J. T. (2000). Effects of *Standards*-based mathematics education: A study of the Core-Plus Mathematics algebra and functions strand. *Journal for Research in Mathematics Education, 31,* 328–361.

This research was supported by a grant from the National Science Foundation (Award #MDR 9255257). The views are those of the authors and do not necessarily reflect those of the Foundation.

bility, and discrete mathematics; organizing topics in a concept-then-skills-then-abstraction order; and reducing attention to formal symbol-manipulation procedures. (For a more detailed overview of the CPMP program, see Hirsch, Coxford, Fey, and Schoen, 1995.)

The purpose of this study was to test the vision of *Standards*-based mathematics education by comparing the effects of the CPMP curriculum and more conventional curricula on growth of student understanding, skill acquisition, and problem-solving ability in algebra and functions.

DESIGN OF THE STUDY

Assessing and comparing students' learning from different curricula are complex problems requiring hard choices among options for data collection and analyses. Assessing knowledge of students with comparable mathematical aptitude and interest prior to curricular treatments makes sense, but in the world of real schools, such ideal samples are not easy to construct. Assessing students on mathematical topics they have had equal opportunity to learn also makes sense, but doing so is difficult when the programs being compared provide different learning opportunities. Timing assessments and finding incentives that will motivate students to attend to testing are complex issues. Another choice researchers must make is whether to study a few students through intensive individual interviews or to study a large number of students through less insightful performance measures.

After careful consideration of these various design options and of the goals of our study, we decided to use a battery of paper-and-pencil instruments to assess the understanding, skill, and problem-solving ability of CPMP and control-group students ending their third year of high school mathematics. We chose the end of Year 3 for testing because at that point in the integrated CPMP curriculum, the core algebra and functions units are completed. We envisioned that the control students would come mostly from advanced algebra classes—again near the end of their high school algebra experiences.

Early in 1997 we invited participants from all 36 high schools that are national field-test sites to participate in this comparative study of algebraic reasoning. Their participation in the study was to be based on three criteria: (a) implementation of the CPMP program with something approximating recommended conditions (heterogeneous grouping, covering the intended curriculum units, and using technology and cooperative learning); (b) identification within the school or a neighboring school of classes using traditional curricula with students of comparable ability; and (c) willingness to devote two class sessions near the end of the school year to testing. Our invitation was accepted at six U.S. schools: two in the Southeast, two in the Midwest, one in the South, and one in the Northwest. At each site were two CPMP teachers and one, two, or three control teachers. The number of students varied from approximately 90 to 180 per site.

As might be expected, establishing comparability of students in the CPMP and control groups was not easy. In four of the sites we were able to obtain standardized mathematics test scores from eighth grade for most students involved in the study. At one of these sites, these data showed that the control and CPMP groups were of comparable ability on entry to high school; at three of these sites we used blocking techniques to construct samples of comparable ability—CPMP students were matched with control-group students who had comparable mathematics achievement or aptitude scores in Grade 8. At the fifth site the students had been randomly assigned to CPMP and control treatments on entry to Grade 9. At the remaining test site we were unable to gain release of eighth-grade test data, but we received repeated assurances that the tested student groups were of comparable ability and just happened to get into different curricular tracks at the start of high school.

We developed three instruments to assess students' understanding, skill, and problem-solving abilities in algebra and functions. Development of these assessment instruments was based on what we considered to be the three main components of effective algebraic thinking: (a) using algebraic ideas and techniques to mathematize quantitative problem situations, (b) using algebraic principles and procedures like solution of equations and inequalities to produce results beyond the information given in the original situation, and (c) interpreting results of mathematical reasoning and calculations in the problematic situation.

Students who have effective command of algebra and functions can execute all three of these processes with skill and understanding. They can use this knowledge to solve significant problems, to discover and confirm important algebraic principles, and to make decisions in situations that depend on quantitative factors. Therefore, we assessed students' performance on comprehensive problems involving all these component processes. But each component of algebraic problem-solving and reasoning activity requires a variety of constituent understandings and skills. Thus we assessed separately these components of algebraic thinking. Finally, because many high-stakes college-admission and placement tests still require skill in algebraic symbol manipulation without the aid of technology, we assessed student performance on that skill with questions devoid of meaningful problem contexts.

In all, we developed three algebra assessments, each with several parallel forms of roughly the same difficulty. The assessments contained more than 100 questions covering a broad range of algebraic ideas, relationships, and techniques. The questions were designed to help us answer two main questions about the effects of the Core-Plus Mathematics curriculum:

- Is the CPMP more effective than a conventional curriculum in developing student ability to solve algebraic problems when those problems are presented in realistic contexts and when students are allowed to use such technological tools as graphing calculators?

- Are conventional curricula more effective than the CPMP in developing students' skills in manipulating symbolic expressions in algebra when those expressions are presented free of application context and when students are not allowed to use tools like graphing calculators?

ASSESSMENTS

Assessment Part 1. Contextualized Problem Solving

In the first assessment (Part 1), we emphasized the type of contextualized problem solving that is typical of CPMP units and other reform curricula.

The students were given 50 minutes to complete Part 1; they were to work individually and were permitted to use scientific or graphing calculators. Each of four parallel forms of Part 1 had four superproblems (a problem setting with five to seven questions about that situation). A portion of a superproblem similar to those in Part 1 is given in Figure 20.1.

Problem 2. The Long-Distance Airliner

Several commercial airlines have nonstop flights from Los Angeles, California, all the way to Sydney, Australia. It is a trip of 7500 miles and can take as long as 18 hours, most of the time over the Pacific Ocean. Therefore, estimating flight time and fuel requirements is very important. One airline uses a formula to predict flight time T in hours from wind speed W in miles per hour.

$$\frac{7500}{500 + W}$$

Answer questions 2.1–2.3 about use of the formula to make safe flight plans. Remember to show your work. If you use a calculator, explain how.

Question 2.1. Find T when $W = -50$ and explain what the result tells about trip flight plans.

Question 2.2. Find the wind speed that will give a flight time of 14 hours.

Question 2.3. Explain the information that will be given by solving the inequality

$$\frac{7500}{500 + W} > 14 \text{ for } W.$$

Figure 20.1. Sample superproblem from Assessment Part 1.

Assessment Part 2. Context-Free Symbolic Manipulations

In the second assessment (Part 2), we emphasized context-free symbolic manipulations that require transforming algebraic expressions and solving equations and systems. The students were given 20 minutes to complete Part 2, and, again, they were to work individually, but this time without using calculators. The two parallel forms of this assessment each comprised six multiple-choice questions and eight constructed-response questions. Two questions typical of those in the Part 2 assessment are given in Figure 20.2.

2. Which of the following expressions is equivalent to $\dfrac{125 + x}{25}$?

   (a) $5x$ (b) $5 + x$ (c) $100 + x$ (d) $\dfrac{126x}{25}$ (e) $5 + \dfrac{x}{25}$

9. Solve the system of equations $\begin{cases} -2x + 3y = 8 \\ x - y = 2. \end{cases}$

   Work Space: *Answer:*

Figure 20.2. Sample items from Assessment Part 2.

Assessment Part 3. Open-Ended Contextual Problems

The third assessment (Part 3) required collaborative work on open-ended contextual problems. The students were paired, and each pair was given 20 minutes to complete the Part 3 assessment; use of scientific or graphing calculators was permitted; graph paper and rulers were provided; and each pair of students submitted one answer paper. Each of the three parallel forms of this assessment consisted of one question encompassing the three main components of effective algebraic thinking described previously. Figure 20.3 shows one of the three problems.

RESULTS

Given the structure of the various assessments, we were able to conduct a variety of analyses in which CPMP students' performance and control-group students' performance were compared and contrasted. We examined first performance on each part of the assessment, and then, in the spirit of "data snooping," we did further analyses, looking for interesting patterns underlying the main effects.

Part 1 assessment items required specific algebraic skills and problem-solving strategies, such as translating problem conditions into symbolic expressions, solving equations, and interpreting results. Development of algebraic ideas through modeling of quantitative relationships in contextual problems is emphasized in the algebra strand of the Core-Plus Mathematics curriculum. CPMP students are also encouraged to make extensive use of graphing calculators as tools for exploring algebraic ideas and solving algebraic problems. Thus, one would expect CPMP students to do better than control stu-

dents in Part 1 of the algebra assessment. As indicated by results in Table 20.1, they did so in our testing.

CPMP students at the end of Course 3 are expected to have some skill in doing symbolic algebra independent of application context and without the use of graphing-calculator technology. Their skills in this area were not expected to be as strong as those of students in a conventional curriculum that included 2 full years of largely symbolic alge-

---

Selling by Telephone

Many companies sell their products through long-distance telephone calls to customers. For example, the *CD Club* sells music compact discs all across the country from its headquarters in New York. Sales calls made by the *CD Club* last an average of about 4 minutes apiece. The club has bids from three possible providers for their long-distance telephone service:

1) *Apple Communications* will charge $0.35 for placing each call and then $0.15 per minute of time used in the call.

2) *Bell Telephone* makes a fixed charge of $260 per week for access to its long-distance lines, but charges only $0.10 per minute of time used by the calls.

3) *Capital Long Distance Services* will make a fixed charge of $600 per week for unlimited use of its long-distance lines.

Question: For the *CD Club* the problem is to choose the long-distance service that is least expensive for their business. What advice would you give about the phone company to choose and why?

Conclusions and Reasoning:

Figure 20.3. Sample problem from Assessment Part 3.

bra, and as shown in Table 20.1, the results from our testing on Part 2 assessment items confirmed that expectation.

The problems on Part 3 required students to integrate specific algebraic skills for work on a complex modeling task. As on Part 1 of the assessment, on Part 3 we expected CPMP students to perform better than students who were of comparable mathematical aptitude and who had experienced a more traditional mathematical curriculum. As shown in Table 20.1, they did so.

Specific Effects of CPMP and Control-Group Curricula

The overall summaries of student performance on applied algebraic problem solving and symbol manipulation show fairly consistent and not surprising differences between the effects of CPMP and those of traditional curricula. In addition to the global analyses on the three parts of the assessment, we performed several additional detailed analyses to illuminate those effects and to suggest ways that curriculum developers can modify materials and teachers can modify implementation of those materials to improve student learning. In the next sections, we examine two aspects of algebra achievement in more detail: (a) algebraic calculation and reasoning and (b) conceptual and procedural knowledge.

Algebraic calculation and reasoning

Formulating an algebraic-function rule, equation, or inequality is only the first step in effective quantitative-problem solving. To draw meaningful conclusions from given information, one invariably needs to perform algebraic calculations, as well— to evaluate expressions, to solve equations and inequalities, and to transform expressions into useful equivalent forms. In traditional curricula, students do the required calculations by following procedural rules for manipulating symbolic expressions. In curricula that make use of numeric, graphic, and symbolic calculating tools, students have available several additional options for answering such questions. Furthermore, in *Standards*-based curricula that make heavy use of real-world contexts for teaching algebraic ideas, students are encouraged to use contextual metaphors as guides to thinking about algebraic tasks.

Many of the specific questions in our Part 1 and Part 2 assessments yielded data for comparison of performance by CPMP and control-group students on algebraic-calculation tasks. These data revealed the following interesting and useful insights into curricular effects:

• CPMP and control-group students demonstrated similar abilities to accurately evaluate linear and quadratic expressions.

• On two items requiring substitution of numerical values into algebraic fractions, performance of control-group students was 30 percentage points higher than that of CPMP students.

• CPMP students generally solved equations more successfully than control-group students did

Table 20.1

Performance on Assessments (All Sites Combined)

	n	Mean	Standard deviation
Part 1. Applied algebra problems		(0–100)	
Control group	273	34.1	14.8
CPMP group	320	42.6	21.3
Part 2. Algebraic symbol manipulation		(0–100)	
Control group	265	38.4	16.2
CPMP group	312	29.0	18.4
Part 3. Open-ended applied problems		(0–4)	
Control group	191	1.07	1.20
CPMP group	184	1.43	1.35

Note. For each part, the difference between the means for the control-group and the CPMP students was statistically significant.

when the equations were embedded in contextual problems.

- Results on solving inequalities revealed a similar pattern of stronger performance by the CPMP students on items stated in a problem context.

The pattern that CPMP students perform better than control-groups students when solving algebraic problems presented in meaningful contexts while having access to calculators and that CPMP students do not perform as well on formal symbol-manipulation tasks without access to context cues or calculators highlights the choices one makes in selecting curricula and teaching methods. The pattern also indicates needed modifications in reform *and* traditional algebra programs.

Conceptual and procedural knowledge

In the design of Part 1 test instruments for our study, we structured the various forms to foster insight into two specific questions about conceptual and procedural knowledge: (a) How will the abilities of CPMP students to *plan* algebraic manipulations compare with their abilities to *do* those calculations accurately? and (b) How will the abilities of CPMP students to *do* algebraic symbol manipulations compare with their abilities to *interpret* the results of those calculations in problem settings? We intended to present some students with contextual problems that required algebraic calculations and to give other students questions that were posed in identical contexts but that required only planning or interpretation of results from symbolic manipulations. The limitations of numbers of participants and available testing time restricted the number of *plan-do* item pairs. However, in 11 pairs of items, we were able to examine the relative difficulties of *doing* and *interpreting* algebraic manipulations. For example, one test form included questions about the economics of a motorcycle business:

> The builders of a new American Eagle motorcycle plan to set a fixed price for the cycle—no rebates or bargaining allowed. The problem is finding the right price to charge. They estimate that their operating costs c will be related to the number n of cycles that are made and sold: $c = 800n + 10{,}032{,}000$.

The students who took one form of the test were asked, "What will the operating cost be if 5000 cycles are made and sold?" The students who took another form of the test were asked to explain briefly the information given by the statement "If $n = 5000$, then $c = 14{,}032{,}000$." They were told not to check the calculations that might be involved in producing that information.

On the calculation items involved in the *do-interpret* pairs of our testing, the performance of the CPMP group and that of the control group were about the same, but the CPMP students generally did better on items that called for interpreting calculated results. The gap between ability to do and ability to interpret algebraic calculations was greater for the control-group students, who, presumably, had much less practice with the interpretive aspects of problem solving. For both the CPMP and the control-group students, doing algebraic calculations was easier than writing interpretations of the results, except on two problems involving graphs.

Taken as a whole and considered with earlier comparisons of performance on problems in context and without context, the data on pairs of items with similar algebraic content but different tasks (*do* or *interpret*) support the notion that learning to interpret the results of algebraic calculations is not highly dependent on ability to perform the calculations. However, even using a curriculum in which greater emphasis is placed on mathematical modeling, including interpreting algebraic calculations in real-world contexts, will not routinely produce students who have mastered that ability.

DISCUSSION AND CONCLUDING REMARKS

Our broad purpose in this study was to test the vision of reform proposals in recent advisory documents, such as the NCTM's *Standards* (1989, 2000) by comparing effects of using a curriculum designed to implement the *Standards* with the effects of using more conventional curricula. We collected and analyzed extensive data on student learning of algebra from both kinds of curricula and found considerable support for main themes of the reform. The most consistent finding of our algebra assessments is perhaps obvious—students learn more about topics that are emphasized in their mathematics classes and less about topics that are not emphasized. We found that students whose algebra teachers emphasize use of functions and graphing technology to solve authentic quantitative problems become more adept at solving such problems than students whose work includes less applied problem solving. We also found that students who devote a great deal of time to practicing symbol-manipulation routines develop greater proficiency at those skills than do students who spend much less time practicing symbolic calculation.

With respect to the traditional processes of algebraic calculation, although students in the CPMP program were not as proficient as the control-group students at manipulating symbolic expressions by hand, they had apparently learned a variety of alternative, calculator-based strategies for accomplishing the same goals. Furthermore, the students who commonly did algebraic calculation in the context of meaningful problem settings developed some proficiency in using those situations as guides to their formal algebra. Whether that use of context cues is a strength or a disability was not clear from our study. Certainly, at some point we want students to be able to deal with algebraic problems free of context cues. Reform curricula that commonly embed algebraic ideas in applied problem-solving explorations may need to better help students abstract and articulate the underlying mathematical ideas. In post-field-test versions of the units, developers have attempted to improve CPMP in this respect.

Whether proponents or opponents of *Standards*-based reforms, mathematics educators looking at the student test scores reported in this study are likely to have the same disappointed reaction we had in scoring the papers. Clearly a great deal of room is left for improvement in the evident achievement of students in algebraic reasoning, problem solving, and calculation. Few students in either the CPMP or the control group could do the kinds of basic symbolic calculation that is common fare on college-admission and placement tests. Even with access to powerful graphing calculators, many students could not accurately solve equations and inequalities. The general level of students' performance on items requiring the mathematization of applied quantitative problems was disappointingly low.

The content of curriculum textbook materials and classroom coverage of those materials make a difference. The question facing those responsible for planning school mathematics curricula is "What mathematics is most important for students to learn?" Our study was not designed to answer that question, but it does show the kinds of trade-offs that might be expected when one allocates time to topics in ways that differ from the allocation in a typical U.S. high school curriculum.

REFERENCES

Hirsch, C., Coxford, A., Fey, J., & Schoen, H. (1995). Teaching sensible mathematics in sense-making ways with the CPMP. *Mathematics Teacher, 88,* 694–700.

National Council of Teachers of Mathematics. (1989). *Curriculum and evaluation standards for school mathematics*. Reston, VA: Author.

National Council of Teachers of Mathematics. (1991). *Professional standards for teaching mathematics*. Reston, VA: Author.

National Council of Teachers of Mathematics. (1995). *Assessment standards for school mathematics*. Reston, VA: Author.

National Council of Teachers of Mathematics. (2000). *Principles and standards for school mathematics*. Reston, VA: Author.

National Research Council. (1989). *Everybody counts: A report to the nation on the future of mathematics education*. Washington, DC: National Academy Press.

GOOD INTENTIONS WERE NOT ENOUGH: LOWER SES STUDENTS' STRUGGLES TO LEARN MATHEMATICS THROUGH PROBLEM SOLVING

Sarah Theule Lubienski, Iowa State University

Abstract. As a researcher-teacher, I examined 7th-graders' experiences with a problem-centered curriculum and pedagogy, focusing on socioeconomic status (SES) differences in students' reactions to learning mathematics through problem solving. Although higher SES students tended to display confidence and solve problems with an eye toward the intended mathematical ideas, lower SES students preferred more external direction and sometimes approached problems in a way that caused them to miss the problems' intended mathematical points. I suggest that class cultural differences could relate to students' approaches to learning mathematics through solving open, contextualized problems.

BECAUSE current reforms in mathematics curriculum and instruction (e.g., National Council of Teachers of Mathematics [NCTM] 1989, 1991, 2000) are intended to help all students gain mathematical power, studying how the reforms affect the least powerful in our society—lower class children—makes sense. Considering class-related equity issues is particularly important for mathematics educators, inasmuch as mathematics often serves as a filter that allows successful students access to high occupational status and pay.

Yet social class is rarely a focus in current educational studies. After reviewing the literature on equity, Secada (1992) noted the lack of serious attention given to social class in mathematics education:

> It is as if social class differences were inevitable or that, if we find them, the results are somehow explained. . . . While the research literature and mathematics-education reform documents (for example, NCTM, 1989; [National Research Council] NRC, 1989) at least mention women and minorities, issues of poverty and social class are absent from their discussions. (p. 640)

Perhaps the educational research community overlooks class-based differences in mathematics achievement because many problems faced by lower class families seem beyond the school's control. Among the barriers that lower class students need to overcome, some, however, arise within schools, even within mathematics classrooms. Changing mathematics curricula and pedagogy can remove or add barriers for such students.

In this chapter I discuss two aspects of reform-inspired mathematics problems that seemed to play out differently in my study for the lower and higher SES students. The first aspect is *openness*, which refers to instructional problems or tasks that have no obvious solution and that can be solved in various ways. The second aspect is *context*, that is, the real-world contexts in which problems were situated.

This chapter is adapted from Lubienski, S. T. (2000). Problem solving as a means toward mathematics for all: An exploratory look through a class lens. *Journal for Research in Mathematics Education, 31,* 454–482.

The Promise of Open, Contextualized Problems

The use of open, contextualized problems seems sensible at many levels. Rather than have students complete meaningless exercises and memorize what the teacher tells them, why not have them learn key mathematical ideas while solving interesting problems? Not only do these ideas seem sensible for all students, but in some ways instruction centered on open, contextualized problems might seem particularly promising for lower SES students.

Some researchers have argued that lower SES students tend to receive more than their share of rote instruction and low-level exercises from teachers with low expectations. This pattern would seem to perpetuate inequalities: Higher SES students are educated to be leaders, whereas lower SES students are trained to be followers. Additionally, according to some scholars, lower class families tend to be more oriented toward contextualized language and meanings.

In this exploratory study I raise questions about the promises of problem-centered teaching while I examine differences in lower SES and higher SES students' experiences in one problem-centered mathematics classroom.

MY STUDY

In this study, I played a dual role as researcher and pilot teacher for a problem-centered curriculum-development project. The study was conducted in a school with a socioeconomic mix of students—a few upper-middle-class students, some middle-class students, some working-class students, and a few lower class students (e.g., with parents with very limited educations, no steady jobs, and incomes below the poverty level).

The school was a pilot site for the Connected Mathematics Project (CMP), a curriculum based on the NCTM Standards and the development of which was funded by the National Science Foundation. The students in this study had used the CMP trial materials during the year preceding this study. While I was a year-long guest in the classroom of a seventh-grade mathematics teacher, my role in the school included pilot-testing the trial materials with one class while serving as a CMP liaison and teaching "model" for other teachers in the school.

Roughly 30 students were in the class (a few came and went across the year). The students were evenly balanced in terms of gender, and, with the exception of two African American males and one Mexican American female (whose family had lived in the United States for several generations), all the students were Caucasian.

Coming from a working-class family myself, I brought to this study an interest in the potential of mathematics education to perpetuate or break the cycle of poverty. I also held expectations that lower SES students *can* succeed in mathematics, as I had succeeded. At the time of this investigation I was completing a doctoral program in which I studied the theories underlying the reforms and worked with key figures in the NCTM reform movement. I was concerned that lower SES students were too often taught to be followers instead of critically thinking leaders, and I was enthusiastic about the reformers' goals of developing reasoning and problem-solving skills in *all* students.

The Teaching and Data-Collection Phase

To get a sense of students' SES backgrounds, I surveyed parents about their occupations, educations, incomes, numbers of books and computers in the homes, and newspapers read regularly: These factors are commonly used SES indicators. I used these data to place the students into two, admittedly rough, categories: lower SES and higher SES. I ultimately gained permission to include 22 of the 30 students, 18 of whom provided clear SES data that allowed me to categorize them as lower SES or higher SES.

I used three sets of interviews, various surveys, student work, teaching-journal entries, and daily audio recordings to document all participating students' experiences, including their struggles and successes with problems, the concepts they thought they were learning, and the factors they found helpful or hindering. A low- and a high-achieving male and female from each SES group were selected as target students to help me distinguish SES from achievement (as measured by students' initial performance in the class, including the quality of their quizzes, homework, and participation).

Survey and interview questions were focused on students' experiences with, and reactions to, the CMP trial curriculum and my pedagogy. Students also completed curriculum surveys regarding the pedagogy and environment in our classroom.

Data Analysis

In analyzing the data I had collected, I focused on similarities and differences in lower SES and higher SES students' experiences. With a primarily Caucasian sample, I was able to focus on class without much variation in race.

Through initial analyses of all the data I had collected, I developed roughly 60 questions or mini-themes, such as reactions to being frustrated on problems, views on whether feelings are hurt in class discussions, and beliefs about which style of instruction helps students think and learn more. For each theme, I created a table in which all relevant survey or interview data for each student were recorded and categorized by SES, then summarized. I paid particular attention to data on high-achieving, lower SES students and low-achieving, higher SES students because doing so helped me sort out differences that seemed more aligned with achievement than with SES. Because many recurring themes related to students' experiences in whole-class discussion, I also systematically examined some of these discussions to compare students' participation with their reported experiences in the discussions.

WHAT I FOUND

During the year, I enjoyed watching some successes, such as the seemingly most apathetic students' becoming engaged in problem explorations, unfold. Yet, in considering which students seemed to experience and benefit from the curriculum in ways the authors intended, I noted some SES-related trends concerning two key aspects of the curricular problems: their openness and their contextualized nature. In the next sections I present data on students' experiences that relate primarily to the open nature of the curricular problems, including students' appreciation of the open problems in relation to more computation-oriented curricula, students' struggles and motivations to make sense of and solve the problems, students' experiences with the problems in real-world contexts, and students' thoughts about what they were learning. (In describing individual reactions, I use one-syllable names for low SES students and three-syllable names for high SES students.)

General Feelings About the Open Nature of the Curriculum

The survey and interview evidence relating to students' curricular preferences indicated that more higher SES students than lower SES students preferred the CMP trial materials to the more typical mathematics books they had experienced through fifth grade. Many students expressed a mixture of opinions about the curriculum, but the four students who consistently expressed preference for CMP were all higher SES. Four of the six students who consistently said that they preferred typical mathematics curriculum to CMP curriculum were lower SES.

In using the CMP materials, students are asked to think harder than, and differently from, before, so I was not surprised that many students reacted negatively to the new expectations. Yet their complaints were not all the same. The predominant theme among the lower SES students was that the problems were frustrating because they were too confusing or hard. For example, Dawn declared, "I don't like this math book because it doesn't explain exactly!" Moreover, six lower SES students specifically said they were better at mathematics as it had been taught previously. For example, James explained, "I'm worse [now], 'cause I used to could do the work, but here I don't understand it." When pushed to explain what exactly was hard or confusing, the lower SES students talked primarily about their difficulty in determining what they were supposed to do with the problems, difficulty they attributed to the vocabulary and general sentence structures used, as well as to the lack of specific directions for how to solve the problems. Because SES correlated with achievement, attention needs to be paid both to those students who were of low SES and high achieving and to students who were of high SES and low achieving. Although both Rose and Lynn were high-achieving students in all subjects, they shared their lower SES peers' attitudes about the difficulty of the curriculum. Rose consistently said that the CMP was harder for her because of the lack of specific direction. She explained, "These books are bad because they are so confusing. We are told to do a page as homework, and the page gives directions, but it doesn't explain how to do it." Lynn said, "I'm better at number problems than problem solving."

Some higher SES students also complained about the books' being confusing, but their complaints were usually voiced toward the beginning of the year (when it was "cool" to complain about the new curriculum) and were not so passionate or personalized. Although no lower SES students said that the CMP curriculum was easier for them than typical curricula, several higher SES students made comments about CMP problems' being easier than computation. Guinevere explained that CMP problems "are a lot easier [for me because] I guess our family's just—we are word-problem kind of people." Benjamin encouraged the CMP authors to make their problems "more challenging."

Internal Versus External Direction

As one might infer from their comments about wanting clearer directions in the CMP trial materials, the lower SES students preferred a more directive teacher role. When asked to rank various modes of working on problems, the lower SES students, especially the girls, ranked "having specific teacher direction" highly. Also, the lower SES girls more often asked me, "Is this right?" Only lower SES students said that they preferred having the teacher "tell them the rules." When the lower SES students talked about aspects of my teaching that were "good," they focused on my ability to explain things well. Teacher direction seemed less important to higher SES students. For example, Andrea liked my "not too strict" style, and Benjamin said, "Ms. Lubienski is a good teacher because she doesn't give answers, but helps and is nice."

In the face of the uncertainty the open problems presented, the lower SES students seemed more passive, less confident about how to proceed. On the various surveys and interviews, lower SES students tended to say that they would become frustrated and give up when "stuck," whereas higher SES students often said that they would think harder about the problem or just interpret it in a sensible way and get on with it.

The higher SES students, more so than the lower SES students, displayed intrinsic motivation to solve the problems and grapple with mathematical ideas. The four students who said that they liked to figure problems out and really understand ideas had higher SES. Although I recorded several examples in my field notes of higher SES students' demonstrating intellectual curiosity and excitement about challenging mathematical problems,

no such examples were noted among lower SES students.

I do not mean to say that the lower SES students never engaged in nor enjoyed the problems, but they seemed to be motivated more by the activities involving fun, games, and contexts of interest to them (such as sports for Nick or dream houses for Rose and Sue); the mathematics did not tend to draw them in or to receive focus. Audiotapes of lower SES students' group work revealed instances of their hurrying through problems but not wanting to appear to be finished for fear I might push them to rethink or discuss their solutions.

Overall, more of the higher SES students than the lower SES students seemed to possess orientations and skills that allowed them to actively interpret the open problems, believe that their interpretations were sensible, and follow their instincts in finding a solution. The lower SES students seemed more concerned with getting clear direction that enabled them to complete their work and were less apt to creatively venture toward a solution.

Contextualization

Most problems in the curriculum were set in some real-world context. A comparison of students' approaches to the contextualized problems revealed a tendency for lower SES students to sometimes approach the problems in ways such that they missed the generalized mathematical point intended by the textbook authors and me. The higher SES students were more likely to approach the problems and discussions with an eye toward the intended, overarching mathematical ideas.

For example, Rose missed the intended mathematical ideas when solving a "find the best buy" problem with volumes and prices of popcorn in three sizes of containers. She had no trouble finding the volumes from the given dimensions, and then she used solid "common sense" reasoning to argue that because the prices went roughly in order of size, the choice should be determined by need: "It depends on how much popcorn you want." Although she had intelligently used the context to determine the degree of accuracy needed, in using this approach, she did not have the intended experience of working with volumes and comparing unit prices. Rose was typical of many of the lower SES students who approached ideas in a contextualized manner, an orientation that sometimes seemed to interfere with those students' understanding of important mathematical ideas. In another example, in a pizza-sharing problem designed to help students learn about fractions, the lower SES students became concerned about many real-world factors, such as who might arrive late to the restaurant and who would want "firsts" and "seconds." These students were very sophisticated in their consideration of multiple, real-world variables, but on their solution paths they did not encounter the intended ideas about fractions.

Other evidence indicates that the lower SES students tended to focus on individual problems without seeing the mathematical ideas connecting various problems. Dawn said that she could never figure out what she was supposed to be learning until she took the test. Rose complained that she had difficulty seeing how the problems we did in class related to the assigned homework problems (which were generally mathematically similar to, but contextually different from, the problems done in class). In contrast, several higher SES students (but no lower SES students) noted that we encountered the same mathematical ideas over and over in various contexts.

Thinking and Learning More

Several students of both SES groups said that the CMP activities made them think more deeply, and most students said that the CMP approach helped them see how mathematics is connected with real life. One difference in the way students talked about the benefits of CMP was that the higher SES students clearly stated that *they* were personally helped by the CMP, whereas the lower SES students talked about its benefits from a more external standpoint—either saying what they thought they *should* be learning or what *others* claimed to be learning.

As with the higher SES students, most lower SES students could articulate much of what the CMP was intending. For example, Rose explained, "CMP is kind of like real-life stuff, so I think that maybe they want you to extend your brain. They aren't going to say, 'Here are the rules'; they want you to figure it out yourself." When pushed to say whether the CMP worked for them as the authors had intended, all higher SES students (except two who gave mixed reviews) said, "Yes." But the lower SES students, especially the girls, tended to be hesitant, saying, "I don't know," or "kind of." Like many other

lower SES students, Lynn spoke in the third person about the curriculum's being helpful. She thought that maybe the other CMP authors and I had a "good reason" for wanting students to figure out the rules themselves, but the lack of clear direction and rules confused her.

A Simple Difference in Previous Achievement, Ability, or Motivation?

Some differences in the students' experiences and views were probably due to previous achievement differences, which correlated with their SES. But various forms of data show that an achievement-differences explanation is incomplete. Rose and Lynn were known as very bright, high achievers in their other, more typical classes (as I learned from their teachers and parents), so why were they struggling with particular aspects of my class? Why did the lower SES students, especially females, tend to say that they understood mathematics better before using the CMP? Why did some higher SES students but no lower SES students say that for them traditional mathematics was more difficult than CMP mathematics?

Many differences in the data might seem attributable to simple motivational differences, but my analyses of the data do not support that hypothesis, particularly when I examine gender and SES together. Gender differences were found in the amount of effort put forth on homework: Girls were more diligent than boys, completing more than 90% of the assignments. The amount of effort put forth on assignments correlated closely with quiz and test results for higher SES males and females and for lower SES males. But the story is quite different for the lower SES females, who showed the most frustration with the class throughout the year. These girls made consistent effort, but they still did not understand the mathematics in a way that allowed them to do well on tests. In other words, effort did not pay off for them in the way it did for the higher SES girls.

Evident in students' perceptions of their own mathematical abilities were patterns that seemed to go beyond actual differences in achievement or ability. In a survey at the end of the year, I asked my students to name the top three mathematics students in the class. No lower SES student ranked himself or herself among the top three—not even Rose, who was considered by nine other students to be a top student. Meanwhile, every higher SES student who was mentioned by anyone else also mentioned himself or herself. Even Samuel and Guinevere, who were not mentioned by anyone else, named themselves. According to their homework and test grades, Rose and Anne had as much cause to feel mathematically confident as Timothy, Guinevere, or Samuel. And yet, they did not.

DISCUSSION

Other potentially confounding factors in addition to achievement need to be addressed. First, one might wonder whether the CMP trial materials or my pedagogy were simply ineffective. Yet students in both SES groups were *differentially* affected by the open, contextualized nature of the problems (and the less directive nature of my pedagogy). Others might wonder whether differences in students' reactions were rooted in their previous school experiences, but inasmuch as the students were all in the same classroom and the majority had been in the same school system since kindergarten, this explanation does not seem plausible.

Because of the nature of this qualitative study, I cannot draw conclusions about the likelihood of differential experiences for both groups of students in other classrooms aligned with current reforms, and I cannot say what would have happened if the students had experienced a reform-inspired curriculum and pedagogy throughout their school years. Additionally, the patterns in the data for this small sample could be due to individual differences that happened to fall along class lines or to other confounding factors, such as teacher-researcher bias.

These concerns prompted me to review the research literature on social-class cultures to shed light on how the patterns in the data might or might not relate to issues of social class. In exploring this literature, I identified ways in which the differences in my students' experiences seem to link with class cultural differences found by other researchers. One major difference I found was students' perseverance and desire for direction when solving problems.

Perseverance and Direction in Problem Solving

The curriculum required students to take initiative in interpreting, exploring, and abstracting mathematical ideas from contextualized problems rather than follow step-by-step rules given at the top of a page. The higher SES students generally felt able to explore the open problems in this way, without becoming overly frustrated. In fact, many high-er SES students voiced appreciation for their increasing mathematical problem-solving abilities. But the lower SES students, especially the females, more consistently complained of feeling confused about what to do with the problems and asked (often passionately) for more teacher direction and a return to typical drill-and-practice problems.

In Heath's (1983) famous study of middle-class and Black and White working-class communities, she found differences in parenting practices. In teaching their children, the middle-class parents emphasized reasoning and discussing. Through interactions with their parents, the children "developed ways of decontextualizing and surrounding with explanatory prose the knowledge gained from selective attention to objects" (p. 56). Meanwhile, the White working-class parents emphasized conformity, giving their children follow-the-number coloring books, for example. They told their children, "Do it like this," for instance, while demonstrating a skill (such as swinging a baseball bat), instead of discussing or explaining the features of the skill or the principles behind it. The White working-class children tried to mimic their parents' actions. When frustrated, the children often tried to "find a way of diverting attention from the task" (p. 62). These children became passive knowledge receivers and did not learn to decontextualize knowledge and then shift it into other contexts or frames. These students did well in early grades, but when they encountered more advanced activities that required more creativity and independence, they frequently had difficulty and asked the teacher, "What do I do here?"

Such comparisons are likely to make many people (including myself) nervous. Bruner (1975), after surveying the literature on social-class cultural differences, concluded, "It's not a simple matter of deficit" (p. 41). But although he could observe strengths and weaknesses in the cultures of both the middle and lower classes, Bruner was particularly concerned about feelings of helplessness and hopelessness that seem more pervasive in the culture of the lower classes. He concluded

I am *not* arguing that middle-class culture is good for all or even good for the middle-class. Indeed, its denial of the problems of dispossession, poverty, and privilege make it contemptible in the eyes of even compassionate critics. . . . But, in effect, insofar as a subculture represents a reaction to defeat and insofar as it is caught by a sense of powerlessness, it [lower class culture] suppresses the potential of those who grow up under its sway by *discouraging problem solving*. The source of powerlessness that such a subculture generates, no matter how moving its by-products, produces instability in the society and unfulfilled promise in human beings. (p. 42, emphasis added)

Taken together, these findings might explain why the lower SES students in this study voiced such strong preferences for a traditional curriculum and teacher role. Those students seemed to have lower self-esteem in the area of their mathematics performance. The lower SES females made strong, consistent efforts to succeed, and yet they struggled when faced with problems requiring more than a standard algorithm. Those females became the most frustrated with the open nature of the problems and my pedagogy, and this finding makes sense, according to the literature, because such students were most likely to internalize their "failure."

Hence, in contrast with the reformers' rhetoric of students' "mathematical empowerment," some of my students reacted to the more open, challenging mathematics problems by becoming overly frustrated and feeling increasingly mathematically disempowered. The lower SES students seemed to prefer more external direction from the textbook and teacher. Those students, particularly the females, seemed to internalize their struggles and "shut down," preferring a more traditional, directive role from the teacher and textbooks. They longed to return to the days in which they could see more direct results for their efforts (e.g., 48 out of 50 correct on the day's worksheet).

IMPLICATIONS

The data and the literature reviewed above could indicate that lower SES students stand to gain the most from problem-solving instruction, inasmuch as it might help these students gain knowledge and skills they are less likely to gain elsewhere. Yet in this study I raise questions about problem solving, not as an end in itself but as a means of learning other mathematical concepts and skills.

From this study of one classroom, one cannot conclude that lower SES students are incapable of functioning in a "reformed" classroom, and one cannot conclude that lower SES students will learn less from problem-centered teaching than from more typical teaching. In fact, some studies have indicated that teaching for meaning can increase economically disadvantaged students' problem-solving and computational abilities when compared with teaching through rote memorization (e.g., Knapp, Shields, & Turnbull, 1995).

I explored various issues that arose in one classroom. More focused studies with larger numbers of students are needed to address the questions raised here. In the meantime, lowering expectations is not the solution for lower SES students. However, this study highlights the dangers of assuming that well-intended changes in instruction will equitably benefit all students. When teachers reform their pedagogy and curricula, they need to support lower SES students' efforts to understand and adapt to new roles. Additionally, these students' learning must be carefully monitored. Through careful analyses of students' writing and talking, teachers must continually assess whether their students are learning mathematics in such a way that they can transfer their understanding from one situation to another.

REFERENCES

Bruner, J. S. (1975). Poverty and childhood. *Oxford Review of Education, 1,* 31–50.

Heath, S. B. (1983). *Ways with words: Language, life, and work in communities and classrooms.* Cambridge, UK: Cambridge University Press.

Knapp, M. S., Shields, P. M., & Turnbull, B. J. (1995). Academic challenge in high-poverty classrooms. *Phi Delta Kappan, 76,* 770–776.

National Council of Teachers of Mathematics. (1989). *Curriculum and evaluation standards for school mathematics.* Reston, VA: Author.

National Council of Teachers of Mathematics. (1991). *Professional standards for teaching mathematics.* Reston, VA: Author.

National Council of Teachers of Mathematics. (2000). *Principles and standards for school Mathematics.* Reston, VA: Author.

Secada, W. G. (1992). Race, ethnicity, social class, language, and achievement in mathematics. In D. A. Grouws (Ed.), *Handbook of research on mathematics teaching and learning* (pp. 623–660). New York: Macmillan.

SECTION IV

RESEARCH RELATED TO ASSESSMENT: INTRODUCTION

SOME authors may be surprised to find their chapters in this section. The chapters were placed here primarily because we chose to focus on the tasks used by the researcher—tasks that could be used by others for assessment purposes. As with other sections, the boundaries are somewhat superficial. After all, we assess learning, we assess teaching, and we assess curricula.

Chapter 22. "A Study of Proof Concepts in Algebra," by Lulu Healy and Celia Hoyles

A study of proof has become a required part of the National Curriculum used in England and Wales. The study of proof in this curriculum is removed from any particular content area, such as geometry. Rather, proof is studied in many contexts. In this study the researchers explored algebra students' notions of proof and ways that those students gained personal conviction of the truth associated with a proof. They examined the characteristics of arguments recognized as proofs by the students and the ways in which the students themselves constructed proofs. This chapter is included in this assessment section because the questionnaire that the researchers devised to explore their research questions is interesting and novel and can be used by other teachers.

In addition to being asked to produce proofs themselves, the students were given a set of quite varied proofs for a familiar statement (when you add any two even numbers, your answer is always even) and for an unfamiliar statement (when you multiply any three consecutive numbers, your answer is always a multiple of 6). They were asked several questions about these proofs—which ones they would use, which ones they thought would receive the highest marks from the teacher,

which proofs they thought were correct, and which ones they found convincing. Some items on this questionnaire were also given to their teachers. One of the many interesting aspects of this study is that students thought that a symbolic proof would garner the most points from the teacher, even though the students themselves would not use that proof.

Lessons learned

In this chapter secondary school teachers will find many relevant lessons concerning students' notions of proof. Do some of these same inconsistencies exist in other countries? Does the fact that the United States does not have a national curriculum make an exploration of students' notions of proof more difficult in this country? Most algebra courses do not include proof. Is this omission a mistake?

The *unintended curriculum* is a term used for learning that takes place without the teacher's conscious knowledge or intent. This chapter presents a prime example of this phenomenon—students came to believe that teachers prefer that algebraic proofs be presented symbolically, even though the teachers did not have this preference. Sometimes, in professional development seminars or courses, teachers are asked to think about, and bring in, examples of the unintended curriculum that they have taught without being aware that they were doing so until asked to reflect on how they teach. Discussions of the examples make other teachers aware that they may have unintentionally been teaching a curriculum that they would not consciously choose; they may, for example, be sending the message that mathematics is a set of rules and procedures to be followed without question and without a need to understand.

Chapter 23. "An Investigation of African American Students' Mathematical Problem Solving," by Carol E. Malloy and M. Gail Jones

The participants in this study were 24 eighth-grade African American students who were enrolled in a university precollege program offered to under-represented students. They were asked to talk aloud while they worked through five problems. They were asked questions after they had solved each problem, and they were interviewed at the end of the session. The researchers coded the problem solutions on a variety of dimensions that have been used in past research on problem solving. They found two characteristics on which African American students differed from mainstream students: These students more frequently chose to use holistic rather than analytic approaches, and they were confident regardless of their success on the problems.

This chapter includes the five problems used with the students. The problems are interesting ones that can be used with other populations. The coding factors are also included here, and readers can find more information on the coding in the original article.

Lessons learned

Whether these results can be replicated is questionable. However, an important realization is that some findings might be different if we look at diverse groups within large groups. As teachers, we must have high expectations for *all* students, but various avenues lead to success, even in mathematics.

Chapter 24. "Grade 6 Students' Preinstructional Use of Equations to Describe and Represent Problem Situations," by Jane O. Swafford and Cynthia W. Langrall

How do students, before receiving any formal instruction in algebra, use equations in problem-solving situations? This chapter explores that question with sixth-grade students. The researchers selected problem situations through which to examine direct variation, linear relationships, arithmetic sequence, exponential relationships, and inverse variation. The students were able to solve problems for specific cases and to describe relationships. They were quite capable of generalizing problem situations, but not necessarily in standard ways. Representing situations symbolically was more difficult for them.

Many teachers will find the carefully designed tasks useful. Not only do they represent different types of mathematical situations, but in each, students had four activities: to solve problems using specific numbers, to describe functional relationships, to represent relationships symbolically, and to write appropriate equations.

Lessons learned

Students in this study were asked for representations, and the reader may be interested in the Brenner et al. study (Chapter 28), in which U.S. students were unable to work with different representations at the same level as their Asian counterparts. The types of problems presented in this chapter will help teachers design their own problems that call for more representational work by their students.

The authors point out other lessons to be learned from this study: Middle school students can profit from more experience with problems involving proportionality, inverse variation, and exponentiations, and they need to examine the same problem through different representations.

Editors' note

Each of the following three brief chapters presents a unique way of measuring students' understanding of function. But do not ignore them if you are not teaching functions. These methods can be adapted to other mathematical topics. Some possible adaptations are discussed in the "Lessons learned" section for each chapter.

Chapter 25. "Using Concept Maps to Assess Conceptual Knowledge," by Carol G. Williams

Concept maps can be used to analyze an individual's organization of his or her knowledge of a particular domain. Such maps have been successfully used in content areas such as biology to find out whether students recognize important relationships, for example, which biological functions are dependent upon other biological functions. In this chapter Williams explores calculus students' understandings of mathematical functions.

In developing a concept map for a particular domain, one first lists all the domain-related terms that he or she can, places each concept in an oval, and links the ovals with lines or arrows. In this study, calculus students and mathematicians were asked to develop concept maps; the mathematicians were told to develop a map of what a calculus student ought to know about functions. Even a glance at the concept maps reveals the intricate connections among concepts as drawn by the mathematicians and the relative lack of connections in the concept maps drawn by students.

Lessons learned

As mathematics teachers, we value the connections among mathematical concepts, and helping students develop those connections is one of our goals. Concept maps can help us determine how successful we have been. Hierarchical concept maps of, for example, geometric shapes or numbers (whole, integer, rational, irrational, real, and even complex numbers) can reveal how students view the relationships existing among the concepts.

Chapter 26. "Students' Conceptual Knowledge of Functions," by Brian O'Callaghan

O'Callaghan presents here a framework for thinking about function knowledge. *Modeling, interpreting,* and *translating* are familiar terms to teachers of algebra, but *reifying* may be a new term. In reification, one recognizes a mathematical process as a mathematical object on which other processes can be performed. Students first think of a function as a process. Thus $y = 3x + 2$ tells the student how to obtain values for y given values for x: The process is to multiply by 3, then add 2. But eventually, teachers hope, a student can begin to think of this expression as a symbol for an object on which one can operate—for example, one can add this function to another function to obtain a third function.

Each of the four components of function knowledge can be measured. In this chapter are examples of tasks that can be used to measure understanding of each component.

Lessons learned

Using this analysis of function knowledge, one can not only create tasks that measure function knowledge but also design curriculum tasks that will lead to comprehensive function knowledge incorporating all four components. Also, the type of careful analysis that was used to develop this framework and its corresponding tasks can serve as an example of how to teach and assess knowledge in other major content domains of mathematics. For example, how could this analysis be applied to the concept of sample? In Chapter 14 the development of understanding of sample is discussed. Could the framework exemplified in this chapter be used to develop a comprehensive analysis of the understanding of sample?

Chapter 27. "Using a Card Sort to Determine One's Understanding of Function," by Gwendolyn M. Lloyd and Melvin (Skip) Wilson

This card sort (see Figure 27.1 in the chapter) was used in the original research study to assess the manner in which a teacher thought about functions. The cards contain graphs, equations, word problems (see Figure 27.2), and tables. They represent linear functions, quadratic and higher degree polynomial functions, exponential functions, logarithmic functions, and trigonometric functions. The cards could be used with second-year algebra students or precalculus students to explore how they think about functions. Diverse ways can be used to sort the cards. Certainly, sorting by representation type would be a start. But then, can the representations be matched? Which aspects of the sort are more difficult than others? Students might undertake this activity in pairs, then compare the ways in which they sorted the cards.

Lessons learned

This card sort was adapted from an earlier one from the book *Developing a Topic Across the Curriculum: Functions*, written by the RADIATE project staff at the University of Georgia and published by Heinemann Press. It could be adapted further to suit any teacher's needs, and the idea of a card sort could be extended to other mathematical topics. For example, one set of cards could have story problems, and others might show calculations. This simple card sort could be used to learn whether students recognize the type of operation that must be used to solve a given word problem. Would the students recognize which story problems call for division? Alternatively, each card could show a different representation of a rational number, including the percent. Does a student match .001 with 1% or with 0.1%? Does a student match 0.4 with 1/4 or with 4/10?

Chapter 28. "Cross-National Comparison of Representational Competence," by Mary E. Brenner, Sally Herman, Hsiu-Zu Ho, and Jules M. Zimmer

These researchers note that significant differences occur in the ways that mathematics is taught in different countries and that consequently students' mathematical competencies might differ. In this study they explore the abilities of students in the United States, China, Taiwan, and Japan to use mathematical representations. The researchers developed a set of *solution questions* that called for either a straightforward computation or a solution to a complex problem. For each solution question, representation questions were written. U.S. students ranked last on solving the solution questions; Chinese students ranked first. Taiwanese and Chinese students ranked high on most of the representation questions; again, U.S. students ranked last. Thus the Asian students had not only stronger basic skills but also stronger conceptual skills than U.S. students had. The authors then looked at other textbook-comparison studies. They found, for example, that Japanese textbooks emphasized coordination of multiple representations, whereas the illustrations in U.S. textbooks did not show linkages among representations. As in the study by Li (Chapter 18), the textbooks used were found to make a difference in what students learned.

Lessons learned

Being able to use and connect representations is a hallmark of understanding mathematics. Assessment tasks are needed to determine how much students know about representations. Being able to create, use, select, apply, and translate among representations is one of the 10 standards in *Principles and Standards for School Mathematics* (National Council of Teachers of Mathematics 2000). When using and reviewing textbooks, teachers need to note the illustrations and decide whether they are helpful. Some textbooks may need to be supplemented to enhance children's learning of representations.

Lessons Learned From Research

CHAPTER 22

A STUDY OF PROOF CONCEPTS IN ALGEBRA

Lulu Healy and Celia Hoyles
Mathematical Sciences, Institute of Education, University of London

Abstract. High-attaining 14- and 15-year-old students simultaneously held two different conceptions of proof in algebra: those about arguments they considered would receive the best mark from the teacher and those about arguments they would adopt for themselves. In the former category, algebraic arguments were popular. In the latter, students preferred arguments that they could evaluate and that they found convincing and explanatory, preferences that excluded algebra. Empirical argument predominated in students' own proof constructions, although most students were aware of its limitations. The most successful students presented proofs in everyday language, not using algebra. Students' responses were influenced mainly by their mathematical competence but also by curricular factors, their views of proof, and their genders.

WITHIN the mathematics community, the topic of proof is frequently the subject of debate; deductive reasoning is contrasted with natural induction from empirical pursuits and with informal argumentation. Yet research indicates that students of mathematics do not find these distinctions easy. The process of proof is undeniably complex, involving a range of student competencies—identifying assumptions, isolating given properties and structures, and organizing logical arguments—each of which, individually, is by no means

trivial. These complexities may be further compounded by the ambiguous nature of the term *proof* itself and by the fact that outside of mathematics, proof can be indistinguishable from evidence.

Within mathematics education, we note that Hanna and Jahnke (1993) suggested that understanding is primary to the acceptance by a learner that a new theorem has been proved, with rigor playing only a secondary role. They went on to argue that students are likely to gain a greater understanding of proof when emphasis is placed on communication of meaning rather than on formal derivation: "A mathematics curriculum which aims to reflect the real role of rigorous proof in mathematics must present it as an indispensable tool in mathematics rather than as the very core of that science" (p. 879).

In mathematics education, empirical research has tended to focus on describing and analyzing students' responses to questions requiring proof. A large body of evidence indicates that most students have difficulties in following or constructing formally presented, deductive arguments; in understanding how they differ from empirical evidence; and in using them to derive further results. Little attention has been paid to documenting students' views of the meaning of proof in mathematics. Additionally, researchers have tended to limit their attention to individual conceptions or to classroom studies, with surprisingly little systematic investigation of school and curricular factors and the role that such factors might play in shaping students' views of, and competencies in, mathematical proof.

The purpose of our study was to analyze how a curriculum, such as the National Curriculum now being used in England and Wales, that allows stu-

This chapter is adapted from Healy, L., & Hoyles, C. (2000). A study of proof conceptions in algebra. *Journal for Research in Mathematics Education, 31,* 396–428.

This project was supported by the Economic and Social Research Council.

dents to test and refine their own conjectures and to gain personal convictions of their truth, affects students' conceptions of proof. This curriculum separates problem solving and proving from mathematical content, particularly from geometry, and is thus in marked contrast with curricula in countries where proof is taught only in traditional Euclidean geometry. The aims of our project were to investigate the characteristics of arguments recognized as proofs by high-attaining students, aged 14–15 years; the reasons behind their judgments; and how they constructed proofs for themselves. We focused on high attainers (around the top 20% of the student population) because these students would have experienced most of the levels of reasoning specified in the National Curriculum. In this chapter we report only on the research we undertook in the domain of algebra.

THE RESEARCH INSTRUMENTS

A questionnaire with three types of items was designed to probe students' views of proof from a variety of standpoints. First, students were asked to provide written descriptions about proof and what they thought it was. Second, students were presented with a mathematical conjecture and a range of different types of arguments in support of it. They were then asked to make a selection from these arguments in accordance with two criteria—which argument would be nearest to their own approach and which they believed would receive the best mark from the teacher. Third, students' assessments of these arguments in terms of their validity or explanatory power were elicited. Two conjectures were included, one familiar (Question A1) and the other unfamiliar (Question A2), and these conjectures, together with the arguments presented in the multiple-choice format, are shown in Figures 22.1 and 22.2.

To provide us with more insight into how the students saw the functions of proof, they were next asked to evaluate each argument presented. They were asked to assess the correctness and generality of each of the arguments. Did they think it contained a mistake? Did they believe that it held for all cases or simply for a specific case or cases? An example of the format used, as it applied to Bonnie's argument in Question A1, is shown in Figure 22.3, Statements 1, 2, and 3. The correctness of students' evaluations of generality was scored by what was called a student's Validity Rating (VR): An entirely correct profile of responses to Statements 1, 2, and

3 for any given argument scored 2; a profile in which the student correctly identified the argument as general, specific, or wrong but was unsure of other factors obtained a rating of 1; and all other profiles scored 0.

The students were also asked to assess to what extent each argument explained the proof and convinced them of its truth, and an example, again relating to the assessment of Bonnie's answer, is also given in Figure 22.3, Statements 4 and 5. If students agreed with both statements, they received 2 points; if they agreed with one or the other of the statements, they received 1 point; otherwise, they received 0. These points represented a rating, called the argument's Explanatory Power (EP).

We sought further to assess the students' feelings for the generality of a proved statement by asking them whether it automatically held for a given subset of cases. The statement that was assumed to have been proved was the familiar conjecture, presented in Question A1, about the sum of two even numbers. The question (A3) posed next is given in Figure 22.4.

The proof questionnaire also included open-format questions for which the students were asked to construct their own proofs, again for one familiar and one less familiar conjecture, and present their arguments so as to obtain the best possible mark (Questions A4 and A5, see Figure 22.5). The order of questions was such that students could adapt arguments presented in previous multiple-choice items for use in their own proof constructions (for example, adapting a proof about the sum of two even numbers to prove a conjecture about the sum of two odd numbers). All the students' constructed proofs were scored for correctness (0 for no basis for proof, 1 for relevant information but no deductions, 2 for partial proof, and 3 for a complete proof). The main form of argument was also classified according to four major categories: empirical, formal (algebraic), narrative, or other. Finally, we coded the students' written descriptions of the role of proof into three categories: truth (verification), explanation (illumination and communication), and discovery (discovery and systemization).

The teachers of the students were also asked to complete all the Questions A1 and A2 in the proof questionnaire, but with a small change in the best-mark criterion. They, like the students, were asked to give their choices of proof and then to choose the

A1.   Arthur, Bonnie, Ceri, Duncan, Eric, and Yvonne were trying to prove whether the following statement is true or false:

When you add any 2 even numbers, your answer is always even.

Arthur's answer

a is any whole number
b is any whole number
2a and 2b are any two even numbers
2a + 2b = 2(a + b)

So Arthur says it's true.

Bonnie's answer

2 + 2 = 4	4 + 2 = 6
2 + 4 = 6	4 + 4 = 8
2 + 6 = 8	4 + 6 = 10

So Bonnie says it's true.

Ceri's answer

Even numbers are numbers that can be divided by 2. When you add numbers with a common factor, 2 in this case, the answer will have the same common factor.

So Ceri says it's true.

Duncan's answer

Even numbers end in 0, 2, 4, 6, or 8. When you add any two of these, the answer will still end in 0, 2, 4, 6, or 8.

So Duncan says it's true.

Eric's answer

Let x = any whole number,
    y = any whole number
x + y = z
z − x = y
z − y = x
z + z − (x + y) = x + y = 2z

So Eric says it's true.

Yvonne's answer

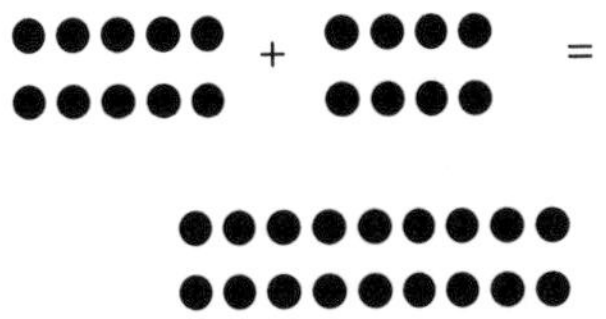

●●●●● + ●●●● =
●●●●●   ●●●●

●●●●●●●●●
●●●●●●●●●

So Yvonne says it's true.

From the above answers, choose one that would be closest to what you would do if you were asked to answer this question.

From the above answers, choose the one to which your teacher would give the best mark.

Figure 22.1. The choices of argument for the familiar conjecture in A1.

proofs that they thought their students would believe would receive the best marks. They were also asked to provide information about their backgrounds, their qualifications, their approaches to teaching proof, and their students' Key Stage 3 (KS3) examination scores (the KS3 tests are national tests administered in the summer term to all Year 9 students in England and Wales).

The student questionnaires were administered to 2,459 students from 94 classes in 90 schools; the mathematics teachers of the 94 classes completed the proof questionnaire and school questionnaire. The sample of 2,459 students was made up of 1305 girls and 1154 boys, all 14 to 15 years of age (Year 10 in England or U.S. Grade 9). These students were of higher-than-average ability, and because the survey was administered toward the end of the school year, all the students had been exposed to the algebra curriculum.

The arguments chosen, the written descriptions given in the proof questionnaire, and the information provided by the teachers were coded, and the

A2. Kate, Leon, Maria, and Nisha were asked to prove whether the following statement is true or false:

When you multiply any 3 consecutive numbers, your answer is always a multiple of 6.

Kate's answer

A multiple of 6 must have factors of 3 and 2.

If you have three consecutive numbers, one will be a multiple of 3, as every third number is in the three times table.

Also, at least one number will be even, and all even numbers are multiples of 2.

If you multiply the three consecutive numbers together, the answer must have at least one factor of 3 and one factor of 2.

So Kate says it's true.

Leon's answer

$1 \times 2 \times 3 = 6$
$2 \times 3 \times 4 = 24$
$4 \times 5 \times 6 = 120$
$6 \times 7 \times 8 = 336$

So Leon says it's true.

Maria's answer

x is any whole number
$x \times (x + 1) \times (x + 2) = (x^2 + 2) \times (x + 2)$
$$= x^3 + x^2 + 2x^2 + 2x$$
Cancelling the xs gives $1 + 1 + 2 + 2 = 6$

So Maria says it's true.

Nisha's answer

Of the three consecutive numbers, the first number is either
EVEN, which can be written $2a$ (a is any whole number), or
ODD, which can be written $2b - 1$ (b is any whole number).

If EVEN
 $2a \times (2a + 1) \times (2a + 2)$ is a multiple of 2
 and either a is a multiple of 3   DONE
  or a is not a multiple of 3
   $\therefore$ $2a$ is not a multiple of 3
   $\therefore$ Either $(2a + 1)$ is a multiple of 3 or $(2a + 2)$ is a multiple of 3 DONE
If ODD
 $(2b - 1) \times 2b \times (2b + 1)$ is a multiple of 2
 and either b is a multiple of 3   DONE
  or b is not a multiple of 3
   $\therefore$ $2b$ is not a multiple of 3
   $\therefore$ Either $(2b - 1)$ is a multiple of 3 or $(2b + 1)$ is a multiple of 3 DONE
So Nisha says it's true.

From the above answers, choose **one** that would be closest to what you would do if you were asked to answer this question.

From the above answers, choose the **one** to which your teacher would give the best mark.

Figure 22.2. The choices of argument for the unfamiliar conjecture in A2.

Lessons Learned From Research

	agree	don't know	disagree
Bonnie's answer			
1. Has a **mistake** in it.	1	2	3
2. Shows that the statement is **always true**.	1	2	3
3. Shows **only** that the statement is true for some even numbers.	1	2	3
4. Shows you **why** the statement is true.	1	2	3
5. Is an easy way to **explain** to someone in your class who is unsure.	1	2	3

Figure 22.3. Assessing the validity and explanatory power of Bonnie's answer.

A3: Suppose it has been proved that when you add any 2 even numbers, your answer is always even.

Zach asks what needs to be done to prove whether when you add 2 even numbers that are square, your answer is always even.

Check either A or B.
(A) Zach doesn't need to do anything; the first statement has already proved this.
(B) Zach needs to construct a new proof.

Figure 22.4. Assessing the generality of a proved statement.

The *familiar* conjecture to be proved was

A4: Prove that **when you add any 2 odd numbers, your answer is always even.** (Write down your answer in the way that would get you the best mark you can.)

The *unfamiliar* conjecture to be proved was

A5: Prove that **if p and q are any two odd numbers, $(p + q) \times (p - q)$ is always a multiple of 4.** (Write your answer in the way that would get you the best mark you can.)

Figure 22.5. The familiar and unfamiliar statements to be proved.

proof constructions were scored. The data were then statistically analyzed.

RESULTS AND DISCUSSION

Choices in the Multiple-Choice Questions

In Table 22.1 we present the distributions of the students' and teachers' choices in the multiple-choice Questions A1 and A2 (see Figures 22.1 and 22.2).

We found a strong statistical difference between the choices the students made for their own approaches and the choices they selected to receive the best score. In fact, the arguments that were the *most* popular for the students' own approaches turned out to be the *least* popular when the students chose an approach to receive the best mark, and vice versa: In answer to the question about the conjecture in A1, Duncan's and Bonnie's arguments were popular for one's own approach but not for the best mark, whereas the reverse was true for Eric's algebraic (but incorrect) attempt and, to a lesser extent, for Arthur's proof.

We also found a statistically significant difference between the teachers' own choices and those that the teachers predicted the students would select for the best mark. In both questions, the most popular among the teachers' choices for their own preferred approaches (Arthur's correct algebraic approach for A1 and Kate's more narrative presentation for A2) were also the most frequently selected by the teachers as the argument that they believed their students would choose for the best mark. Both sets of data indicate that students judged that their teachers would reward any argument provided it contained some "algebra," where-

Table 22.1

Distribution of Students' and Teachers' Choices of Proofs for Familiar Conjecture A1 and for Unfamiliar Conjecture A2

Argument	Percentages of students		Percentages of teachers	
	Own approach	Best mark	Own approach	Best mark
Argument chosen for A1	*N* = 2450	*N* = 2423	*N* = 94	*N* = 94
Duncan (narrative)	29	7	6	12
Bonnie (empirical)	24	3	3	7
Ceri (narrative)	17	18	10	11
Yvonne (visual)	16	9	—	—
Arthur (algebraic)	12	22	81	62
Eric (algebraic)	2	42	0	9
Argument chosen for A2	*N* = 2381	*N* = 2348	*N* = 94	*N* = 94
Kate (narrative)	41	19	70	48
Leon (empirical)	39	2	4	7
Nisha (algebraic)	7	55	22	38
Maria (algebraic)	13	24	3	6

Note. Yvonne's response was not given to the teachers.

as teachers presumed that the logic of the argument would also be important. The teachers thus appeared to overestimate the extent to which their students would make judgments that were based on mathematical content rather than simply on form.

Students' Constructed-Proof Scores

After scoring all the students' proofs (see Figure 22.5), we compared the distribution of the scores with the distribution of choices of correct proofs in the multiple-choice questions. A comparison of the total number of students who selected an argument representing what we deemed to be a correct proof with the total number of students who constructed either a partial or complete proof showed that students were significantly better at choosing correct mathematical proofs than at constructing them. In fact, the students were rather poor at constructing proofs, as shown in Table 22.2, which presents the distributions and means of students' scores for the

proofs to the familiar and unfamiliar Conjectures A4 and A5.

Not surprisingly, the students constructed better arguments for the familiar conjecture than for the unfamiliar one, with 40% using some deductive reasoning in the former case (22% completely correct together with 18% partial proofs). For the less familiar and more complex statement, more than a third of the student sample could not give any basis for a proof, and only 3% managed to produce a complete proof.

Clearly, constructing a proof was difficult for students. To obtain more insight into what students believed a proof should look like, we analyzed the major forms in which their arguments were presented. Our analysis of students' choices of argument had pointed to their preferences for empirical, narrative, and formal (algebraic) proofs, the first two for a student's own approach and the last for the best mark. We therefore focused on the distribution (shown in

Table 22.2

Distribution and Mean of Students' Scores for Each Constructed Proof

Constructed-proof score	Familiar conjecture (A4)[a]		Unfamiliar conjecture (A5)[b]	
	No.	%	No.	%
0 No basis for the construction of a correct proof	354	14	866	35
1 No deductions but relevant information presented	1130	46	1356	55
2 Partial proof, including all information needed but some reasoning omitted	438	18	154	6
3 Complete proof	537	22	83	3

Note. Proofs were scored 0 for no basis, 1 for relevant information but no deductions, 2 for partial proof, and 3 for correct proof.

[a]*M* = 1.47; *SD* = 0.988. [b]*M* = 0.778; *SD* = 0.708.

Table 22.3
Distribution of Forms of Presentation for Constructed Proofs

Form of proof	Familiar conjecture (A4)		Unfamiliar conjecture (A5)	
	No.	%	No.	%
Empirical	845	34	1062	43
Narrative	692	28	792	32
Formal (algebraic)	281	11	82	3
None	74	3	443	18
Other	567	24[a]	80	4

[a]Eight percent of the responses for A4 were attempts at visual proofs, and 15% were attempts to produce an exhaustive proof by examples referring to the units digit. These types of proof did not appear in responses to A5.

Table 22.3) of these major forms of argument among the students' own productions.

Table 22.3 shows that although empirical example was the most popular form of argument used by the students, if they did try to go beyond this pragmatic approach, students were more likely to give arguments expressed informally in a narrative style than to use algebra formally. These narrative proofs clearly were more likely than the algebraic attempts to be correct.

Although arguments that included algebra were the most popular among students for the best mark, our results show that students knew that they would be highly unlikely to base their own arguments on similar algebraic constructions. In both multiple-choice questions (A1 and A2 in Figures 22.1 and 22.2 and Table 22.1), the algebraic arguments were the least frequently selected by the students as being closest to the approach they would use, and algebra was rarely used as the language through which the students attempted to write their own proofs (11% for Conjecture A4, 3% for Conjecture A5; see Figure 22.5 and Table 22.3). When the students did give algebraic arguments, frequently the arguments scored 0.

Students' Views of the Role and Generality of Proof

The students' views of proof were evident in their choices of proofs, their evaluations of those choices, and in their own proofs—although the students' constructions were also influenced by their understanding of the mathematics involved. We acquired further evidence about their views of the generality of a proof through their responses to Question A3 (See Figure 22.4). Among our student sample, the majority of students (62%) were aware

that no further work was necessary to check whether a proof held for a subset of cases if its generality had already been proved.

Factors Influencing Students' Conceptions of Proof

Using a series of statistical models with school variables, we found that only two variables—the gender of the student and his or her test score on the required Grade 9 KS3 Test—were significantly associated with both the choices made on Questions A1 and A2 and the scores for Conjectures A4 and A5. In terms of gender, we found that when we accounted for KS3 test scores, girls obtained higher scores than boys in their constructed proofs.

The students with high KS3 test scores constructed better proofs than those with low KS3 scores and were less likely to rely only on empirical evidence in their constructions and selections. This result in itself is not altogether surprising, although it should be noted that KS3 tests include no items on proof. Perhaps more interesting is the fact that the KS3 test score was *never* the only factor associated with student performance, and other factors exerted significant influence. Other factors, such as having some idea of the nature of proof or being in a class preparing for the most challenging national examination, could also be of influence.

The students' choices in Questions A1 and A2 indicated that they were attracted in their choices for their own approaches to arguments that they could evaluate correctly (measured by their VRs, i.e., finding mistakes or correctly assessing their generality) and those that they felt were explanatory (explanatory power was a significant variable). The students' choices were also influenced by their views of proof and its role and by which arguments they believed the teacher would reward with the best mark.

Overall the most interesting trends relating to students' choices were that increases in KS3 test scores and hours per week of mathematics class consistently raised the likelihood of choosing arguments that were not empirical, *except* in the case of Eric's algebraic but incorrect argument. A situation in which students have high mathematics attainment and considerable exposure to mathematics might be associated with students' choices of proofs that are more mathematical but does not guard against their attraction to *x*s and *y*s. The finding that if all other factors were taken into account, girls were less likely than boys to choose Arthur's argument may also be worthy of further investigation.

Ratings of validity and explanatory power were significantly associated with students' choices for their own approaches for almost all the arguments presented. A consistent and positive effect showed that the higher these scores for any argument, the greater the chance that the argument would be selected.

Perhaps the most surprising finding was that no variation was found in students' scores according to the teacher variables of qualifications, gender, and teaching experience, although almost all the teachers in the sample were well qualified mathematically.

We had set out to investigate school variation, but after adjusting for all the significant factors, we found considerable overlap in constructed-proof scores between schools and, in particular, more unexplained variation within than between schools. We also found that little variation remained at the school level after adjustment had been made for all the significant variables.

SUMMARY AND IMPLICATIONS

Our study showed that the majority of the students were unable to construct valid proofs in the domain of algebra, but it also indicated that they valued general and explanatory arguments. Additionally, although students predominantly used empirical arguments for their own proofs, they also recognized that these arguments had low status and would not receive the highest marks from their teachers. The majority were also aware that empirical arguments were not general—particularly if the statement to be proved was not familiar—but recognized that examples offered a powerful means of gaining conviction about a statement's truth. Most students were also aware that a valid proof must be general and that once a proof has been given, no further work is necessary to ascertain the truth of specific cases within its domain of validity.

The students were reasonably successful at evaluating narrative arguments and were likely to see them as explanatory. Also, students had most success in constructing their own proofs when they used this narrative form. In contrast, the students found that arguments containing algebra were hard to follow and offered little in terms of communicating and explaining the mathematics involved. Students still believed that the use of complicated algebra expressions would get the best marks from their teachers—a tendency about which the teachers seemed largely unaware. Yet few students chose such arguments for their own approaches, and fewer still constructed them with any success.

The fact that students rarely used algebra in their own proofs raises an important question. If students do not see algebra as a language with which they can explain phenomena in mathematics classrooms in which explanations are highly valued, and if they can successfully construct informal arguments, how can they be motivated to re-express their arguments algebraically? Clearly, such a question extends beyond notions of proof to encompass students' views of algebra. Our students had yet to see algebraic transformations as potential sources for conjectures, as building blocks for new proofs, or as a means to explain and communicate their mathematical ideas.

Teachers now need to find ways to build on the particular conceptions of our students. We need to exploit their strengths in informal argumentation and narrative proof so that they develop more multifaceted competencies in their views of proof and in proving.

REFERENCE

Hanna, G., & Jahnke, H. N. (Eds.). (1993). Aspects of proof [Special Issue]. *Educational Studies in Mathematics, 24* (4).

CHAPTER 23

AN INVESTIGATION OF AFRICAN AMERICAN STUDENTS' MATHEMATICAL PROBLEM SOLVING

Carol E. Malloy and M. Gail Jones
University of North Carolina at Chapel Hill

Abstract. In a study of the problem-solving characteristics, strategy selection and use, and answer-verification actions of 24 African American 8th-grade students, the students displayed approaches attributed to African American learners in the literature: They regularly used holistic rather than analytic reasoning, and their display of confidence and high self-esteem did not appear to be related to success. Successful students' problem-solving actions matched previously reported actions of good mathematical problem solvers: successful use of strategies, flexibility in approach, use of verification-of-solution actions, and success in dealing with irrelevant detail.

MOTIVATED by the lack of available empirical research about how African American students solve mathematics problems and by the uneven achievement reports for these students, we investigated how 24 African American students approached and solved five mathematics problems and how these students perceived themselves as mathematics students. In studying problem solutions, we were specifically interested in the actions that contributed to success.

Although many researchers have studied mathematical problem solvers, few have focused on mathematical problem solving by African American students. But studies of learning preferences indicate that African American students' approaches to learning may be characterized by factors of social

and affective emphasis, harmony with their communities, holistic perspectives, expressive creativity, and nonverbal communication (Hale-Benson, 1986; Willis, 1992). Research indicates that African American students are flexible and open-minded rather than structured in their perceptions of ideas (Shade, 1992).

In this study we sought to describe how a particular group of African American students solved mathematics problems. We began this study with the premise that many African American students were successful in mathematics, and we investigated both cognitive and affective aspects of the mathematical problem-solving processes of our students. Our question was "How do these African American eighth-grade students solve mathematics problems?" Our broad goal was to determine their skills, the levels of their understanding of the problems, the processes they used to solve the problems, their attitudes about their mathematics abilities, and the effects of their culture on their mathematics learning.

Twenty-four Grade 8 students participated in the study. All were enrolled in a university-sponsored precollege program designed to offer underrepresented middle school and high school students enrichment, encouragement, and career exploration in mathematics and science through summer and academic-year activities. The sample population reflected the population in the precollege program.

Each of 24 students participated in an individual talk-aloud problem-solving session during which the student was observed solving five mathematics problems. The session had four features: (a) a

This chapter is adapted from Malloy, C. E., & Jones, M. G. (1998). An investigation of African American students' mathematical problem solving. *Journal for Research in Mathematics Education 29*, 143–163.

warm-up problem and review; (b) five think-aloud problems to solve, one at a time; (c) retrospective interviews after the completion of each problem; and (d) a follow-up interview at the end of the session. The sessions lasted 60 to 75 minutes. The students talked aloud about their thoughts while they worked on each of the five problems separately. The students were allowed 15 minutes for Problem 4 and 10 minutes for each of the other four problems. After completing initial work on each problem, the student was asked the same 12 questions about the process she or he had used to solve the problem, including "How did you feel about the question?" "Could you solve the problem in another way?" "What made you decide to solve the problem the way you did?" and "Did you check your answer?" After the student had completed the five problems, she or he was asked 14 questions, including "What does it mean to have an understanding of mathematics?" "How important is it to get the right answer?" "How do you start to solve a problem if you don't know how to find the answer?" and "How important is the approval of the teacher to your achievement?"

THE PROBLEMS

The problems (see Figure 23.1) were selected using the following criteria: (a) Each problem could be solved using multiple strategies; (b) the problems selected varied in difficulty; (c) each problem could be solved using holistic or analytic approaches; and (d) each problem required inferential, deductive, or inductive reasoning.

The categories of actions and the approaches that emerged from the transcriptions of the interviews were coded as follows: (a) *holistic* or *analytic* (worked from the whole to the solution or broke the problem into related parts and worked toward the solution); (b) *linear* or *flexible* (used one solution plan or tried varied plans); (c) *meaningful* or *superficial* (used purposeful solution method or solution method unrelated to the problem); (d) *persistent* or *erratic* (used appropriate methods over allotted time to solve problems or either withdrew or used irrational mathematical computations); (e) *confident* or *not confident* (said "I'm sure this is correct" or "I'm not sure of this problem"); and (f) *conceptual* or *procedural* (understood the mathematical ideas behind the problem or approached the problem algorithmically). Thus, the approaches students took toward understanding a problem, the types of strategies they used, the manner in

which they organized their work, and the manner in which they evaluated their solutions were studied.

RESULTS

The cognitive aspects of the students' problem solving that emerged from the analyses included their flexibility in implementing solution plans and their use of holistic and analytic reasoning. The noncognitive aspects that emerged were the students' persistence and desire for understanding, their self-confidence, and their self-esteem. Each aspect is discussed through the students' problem solutions.

Flexibility

Flexibility, that is, the willingness to think about and use diverse solutions, is a characteristic of good problem solvers (Hembree, 1992). Successful students in this study demonstrated their flexibility in three phases of problem solving: choosing a plan, implementing that plan, and selecting a strategy. If the successful students had not solved the problem after they had moved through the orientation, organization, execution, and verification of their initial plans, they reconsidered their actions by rereading the problem and devising alternative plans and strategies to find the correct solution. Successful students monitored their actions, whereas unsuccessful students did not. For example, Davon, a successful student, explained his correct solution while he used a calculator.

Davon: How can I divide 6 into 1500? Two hundred fifty. See the difference; I [am] trying to see the difference between the pounds of hamburger; 375 minus 250 equals 125 pounds difference. (Pause)

Researcher: What are you thinking?

Davon: I am trying to figure out if I have the same amount of packages of buns. Actually I wouldn't need to know that. I'll try to divide 375 by 6. That is 62.5 pounds of hamburger each day. So if he had a decrease on the pounds during the weekday, see how many times 62.5 would go into 250. I think it would last 4 days.

Davon used four approaches, starting with the number of hamburgers, moving through the difference in poundage and packages of buns, and, finally, moving to the daily rate.

Car Wash

Nakisha, Gregory, Kerstin, and Brandon had a car wash on Saturday. Nakisha washed twice as many cars as Gregory. Gregory washed 1 fewer than Kerstin. Kerstin washed 6 more than Brandon. Brandon washed 6 cars. How many cars did Nakisha wash?

Marbles

Marbles are arranged as follows:

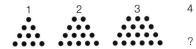

How many marbles would be in the fourth arrangement?
How many marbles would be in the 25th arrangement?

Hamburger

Henry's Hamburger Heaven is open 6 days a week and sells an average of 1,500 hamburgers a week. Each week, 375 pounds of hamburger and 125 packages of buns are used. The supplier brought 250 pounds of hamburger on Wednesday. How many days would this hamburger last?

Church

At a community church, the leader plans to place the page numbers for three different songs on a board in the front of the church. The leader must buy plastic cards to put on the board. Each card has one large digit on it. The leader wants to buy as few cards as possible. The song book has songs numbered from 1 to 632. What is the fewest number of cards that must be purchased to make sure that it is possible to display any selection of three different songs?

Triangle

In the three-sided figure below, the number in each square must be equal to the sum of the numbers in the two circles on either side of the square. Find the numbers that go inside the circles.

Answers: 1) 22; 2) 22, 106; 3) 4 days; 4) 65 tiles; 5) 8, 6, 11

Figure 23.1. Problems.

Holistic and analytic reasoning

Students used holistic- or analytic-reasoning approaches alone or in combination when they solved the problems. In her solution to the Church Problem, Denise approached the problem in a holistic manner, using synthesis to narrow the paths to the solution. She began by determining the maximum number of tiles needed to place the three song numbers on the board (90 tiles). Then she began to synthesize the information to reach a solution that included songs numbered only to 632. Denise concluded, "Okay. I think the answer is 65." In 10 minutes Denise started from the whole and combined the information to find the correct answer; she used a holistic approach. In contrast, an analytic approach to this problem might be to investigate the number of tiles needed for each digit and then to find the total number of tiles needed for the entire set. Most students used a holistic approach on at least one occasion. In 75% of these cases, the students were successful.

Persistence and desire for understanding

Students' attitudes were similar across degrees of success at problem solving. Their persistence was reflected in the desire to be successful in mathematics. Most students did not allow frustration with difficult problems to influence their persistence in seeking a solution. Most remained confidently engaged for the full 10 or 15 minutes, determined not to give up on the problems. If they could not find the solution quickly and if they did not have a plan for the solution, they would reread the problem statement and try to find a clue to assist them. This persistence helped students to reach the "Aha!" experience that often resulted in a correct solution. In the interviews, most students stated that they thought of their mistakes as learning experiences. They said that when they got a wrong answer to a problem, they would typically do the problem over until it was right.

When students were asked to describe what made them successful, they said that the most important factor in their success was understanding. They knew that they needed to have both conceptual understanding and procedural skills to understand mathematics. In an interview, Theresa, a student who valued both, said of understanding,

> To know how to do it and how it works, and stuff like that. . . . I understand how the radius and diameter work, but I don't understand *it*. I can work a problem, but I don't understand what it is and what it really means. . . . I don't understand.

Theresa was not satisfied to know only how to solve the problem; she wanted to understand the mathematical concepts underlying the procedures.

Confidence and self-esteem

Research has shown that confidence and self-esteem are related to problem-solving success, but this study's findings were not consistent with that research. As determined from the students' comments in retrospective interviews, the students were confident that their work was correct in 65% of their problem solutions. On the whole, 20 of the 24 were confident about the majority of their work on the five problems. Even students who used erratic solution plans often stated that their solutions were reasonable.

Students' confidence in their mathematics abilities and in their problem solutions was demonstrated in several ways. Students expressed confidence in their solutions; some chose not to verify solutions because they were sure that their answers were correct; some chose to verify their work to ensure correct answers. In follow-up interviews, students reported that they saw themselves as "doers" of mathematics.

In this study we found two problem-solving characteristics related to learning preferences of African Americans to be different from characteristics found in the mainstream population: students' frequent use of holistic reasoning and students' confidence in their mathematical abilities, regardless of success. The differences are in part cognitive, but they also address the students' affective realms—the students' attitudes about themselves, their needs, and their motivations for learning. These factors influence students' cognitive and affective development, their worlds, their thinking, and therefore their cultures of learning.

The students in this study showed that they could use *both* holistic and analytic reasoning to solve problems, but they used holistic reasoning more frequently. African American parents instruct their children to view their worlds and environments in totality, because African American children's safety and success often depend upon cognizance of their surroundings. A natural result is that African American people tend to respond to experiences holistically, in terms of the whole picture instead of its parts (Willis, 1992). They tend to prefer holistic and inferential reasoning to analytic reasoning.

Students' confidence and self-esteem may have resulted from prior successes in mathematics and from repeated praise and encouragement from parents, friends, and neighbors. Another contributing factor may have been their membership in the pre-college program. These students had participated in communal and institutional networks in which they were socialized to recognize their group and personal responsibilities. They had learned that confidence and self-esteem, as attributes of resilience, can mediate feelings of vulnerability perceived when institutional expectations for achievement are low. The success of these students may have been a reflection of their knowledge that they were in control of their achievement.

However, half the 24 students in this study, similar to many African American learners in mainstream classrooms, showed inability to adequately solve nonroutine problems. We are left with the question "How can the more positive characteristics demonstrated in this group's performance be achieved in everyday settings?"

Educational research has given marginal status to factors that contribute to mathematics achievement in diverse populations. Mathematics reform as it relates to nonmainstream students has too often been predicated on research and interventions that begin with the assumption that these populations historically have had low achievement (Secada, 1991). Those students who have learned in spite of social forces, reduced opportunities, and low expectations are considered aberrations. This investigation belies such assumptions. Educators must not agree to accept lower expectations from African American students. The challenge is rather to find the means to reach African American students, to enable them to achieve success to the same degree as that of their mainstream classmates.

REFERENCES

Hale-Benson, J. (1986). *African-American children: Their roots, culture, and learning styles*. Baltimore: Johns Hopkins Press.

Hembree, R. (1992). Experiments and relational studies in problem solving: A meta-analysis. *Journal for Research in Mathematics Education, 23*, 242–273.

Secada, W. (1991). Diversity, equity, and cognitivist research. In E. Fennema, T. P. Carpenter, & S. J. Lamon (Eds.), *Integrating research on teaching and learning mathematics* (pp. 17–53). Albany: SUNY.

Shade, B. (1992). Is there an Afro-American cognitive style? An exploratory study. In A. Burlew, W. C. Banks, H. McAdoo, & D. Azibo (Eds.), *African-American psychology* (pp. 256–259). Newbury Park, CA: Sage.

Willis, M. (1992). Learning styles of African-American children: Review of the literature and interventions. In A. Burlew, W. Banks, H. McAdoo, & D. Azibo (Eds.), *African-American psychology* (pp. 260–278). Newbury Park, CA: Sage.

GRADE 6 STUDENTS' PREINSTRUCTIONAL USE OF EQUATIONS TO DESCRIBE AND REPRESENT PROBLEM SITUATIONS

Jane O. Swafford and Cynthia W. Langrall, Illinois State University

Abstract. The purpose of this study was to investigate 6th-grade students' use of equations to describe and represent problem situations prior to formal instruction in algebra. Ten students were presented with a series of similar tasks in 6 problem contexts representing linear and nonlinear situations. The students in this study showed remarkable ability to generalize the problem situations and to write equations using variables, often in non-standard form, but they rarely used their equations to solve related problems. Students' preinstructional uses of equations to generalize problem situations are described and questions about the most appropriate curriculum for building on students' intuitive knowledge of algebra are raised.

BECAUSE of advancements in the use of technology and its prevalence in our culture, greater understanding of the fundamentals of algebra and algebraic reasoning is necessary for all members of society. Accordingly, the National Council of Teachers of Mathematics (NCTM) has recommended that algebra be studied by all students, including those who are low achieving or underserved (Edwards, 1990). Further, the NCTM (1991) recommended that mathematics instruction be built on students' informal or prior knowledge. Although extensive research on algebra learning has been conducted, educators do not

have a complete picture of what students can do in algebra prior to formal instruction. Thus, they do not know which aspects of informal knowledge create a useful foundation upon which instruction can be built.

Our overarching goal for this study was to determine the extent of students' use of equations to describe and represent contextual problem situations prior to their formal study of algebra. We expected that prior to instruction in algebra, students would be able to find answers by performing numerical operations on given values of the independent variable or on previously computed values of the dependent variable.

Next, we wanted to determine whether the students would be able to generalize the relationship. To generalize a problem situation is to identify the operators and sequence of operations that are common among the specific cases and extend them to the general case. A generalization of a problem situation may be presented verbally or symbolically using variables. Either of these approaches would be considered a representation of the problem situation. A representation of a mathematical problem situation is a depiction of the relationships and operations in the situation. Representations can take a variety of forms from verbal to symbolic and can represent specific cases or the general case. Diagrams are pictorial representations of one or more specific cases, and tables are systematic representations of a series of specific cases. Graphical representations can, if viewed pointwise, represent specific cases and, if viewed holistically, represent generalizations. Narrative descriptions of the general case are verbal representations of the generaliza-

This chapter is adapted from Swafford, J. O., & Langrall, C. W. (2000). Grade 6 students' preinstructional use of equations to describe and represent problem situations. *Journal for Research in Mathematics Education, 31*, 89–112.

tion, whereas equations using variables are symbolic representations of the general case.

With respect to representations, we first wanted to determine whether students could verbally describe the general case. Next we asked students to use variables to symbolically represent the relationship in the problem situation. We wanted to know whether students could produce symbolic representations prior to instruction in algebra and, if so, what forms they would take. Finally, we wanted to determine whether and how students would use their symbolic representations to solve related problems. That is, prior to instruction in algebra, do students treat equations as mathematical objects in their own right and operate on them using algebraic manipulations?

We decided to pose the same set of tasks over different mathematical domains and for different problem types within the class of linear functions. By examining students' responses to the same tasks across a variety of mathematical domains, we hoped to be able to give a robust description of their preinstructional use of equations to generalize problem situations. Further, by contrasting students' performance across different families of functions and within the family of linear functions, we hoped to contribute useful insights upon which instruction could be built.

We selected five boys and five girls from a Grade 6 classroom; none had received any formal instruction in algebra. We interviewed each student individually, presenting the same six verbal problem situations and asking the students to solve similar tasks for each (see Figure 24.1). The situations involved familiar contexts and represented the following mathematical content domains: direct variation/proportionality, linear relationships, arithmetic sequence, exponential relationships, and inverse variation. For each situation, the students were first asked to solve problems involving specific numbers. Then they were asked to describe the functional relationship between the independent and dependent variables and to write a general equation. If they had difficulty describing a functional relationship, the students were asked to construct a table or were shown a completed table if they had not previously made one; they were asked whether they saw any relationships between the numbers on the left side and the numbers on the right side of the table. Finally, we posed related questions that could be answered about the situation by using the equation, either by solving for specific values or by substitution. Each interview lasted approximately 45 minutes and was audiotaped and later transcribed.

WHAT WE FOUND

The sixth-grade students in this study showed remarkable ability to generalize problem situations by describing relationships and writing appropriate equations using variables, although their notation was sometimes nonstandard. However, more of the students were able to describe the relationships than were able to represent them symbolically. Even though the students were often able to write equations, few used them to solve related problems, and those who did so used their equations as a list of operations to be performed. A summary of their successes can be found in Figure 24.2.

Computing specific cases

Most of the students were able to solve problems involving specific cases, with the exception of the Car Wash Problem (inverse variation). For the Refund Problem, Wage Problem, and Border Problem, students used the functional relationships between the independent and dependent variables, implicit in the problems, to solve specific cases. For the Concert Hall Problem and Paper Folding Problem, the students tended to compute solutions to specific cases by using the recursive relationship described in the problem. For the Concert Hall Problem, they would begin with Row 1 and add 2 for each successive row until they got to the desired row. Similarly with the Paper Folding Problem, the students would continue doubling the number of regions obtained with each fold until they reached the desired number of folds. In the Car Wash Problem, which is an inverse-variation situation, the students could compute the time required to wash the cars given half or twice as many washers, but they had difficulty computing the time required beyond these cases.

Describing relations

The students' abilities to verbally describe the relationships in the six items tended to mirror their performance in solving specific cases. Once again for the Refund Problem, the Wage Problem, and the Border Problem, most students correctly described the functional relationships between the independent and dependent variables represented in the

Lessons Learned From Research

Refund Problem (direct variation/proportionality)
In some states, a deposit is charged on aluminum pop cans and is refunded when the cans are returned. In New York, the deposit is 5 cents a can.
a) What would be the refund for returning 6 (10 or 12) cans?
b) Describe how the store owner would figure the amount of refund for any number of returned cans.
c) Let *R* represent the amount of refund, and let *C* represent the number of cans returned; write an equation for the amount of refund.
d) Can you use your equation to find out how many cans would have to be returned to get a refund of $3.00? How much refund would you get for 100 cans?

Hours and Wage Problem (linear relationship)
Mary's basic wage is $20 per week. She is also paid another $2 for each hour of overtime she works.
a) What would her total wage be if she worked 4 hours of overtime in 1 week? 10 hours of overtime?
b) Describe the relationship between the amount of hours overtime Mary works and her total wage.
c) If *H* stands for the number of hours of overtime Mary works and if *W* stands for her total wage, write an equation for finding Mary's total wage.
d) Can you use your equation to find out how much overtime Mary would have to work to earn a total wage of $50? One week Mary earned $36. A coworker of Mary's had worked 1 hour less overtime than Mary had worked. What would her friend's wage be for the week?

Border Problem (linear relationship, geometric context)
Here is a 10 by 10 grid. How many squares are in the border?
a) Here is a 5 by 5 grid. How many squares are in the border of this grid? We don't have a 100 by 100 grid, but how could you figure out how many squares would be in the border?
b) Describe how to figure out the number of squares in the border of any size grid (*N* by *N*).
c) Let *N* represent the number of squares along one edge of a grid, and let *B* represent the number of squares in the border. Write an equation for finding the number of squares in the border.
d) Can you use your equation to find the size of a grid with a border that contains 76 squares?

Concert Hall Problem (arithmetic sequence)
The first row of a concert hall has 10 seats. Each row thereafter has 2 more seats than the row in front of it.
a) How many seats are in Row 10? (If student had difficulty, ask how many seats in Rows 2, 3, and 4.)
b) The ticket manager needs to know how many seats are in each row. If she knows the number of the row, explain how she can figure out how many seats are in that row.
c) Let *R* represent the number of the row, and let *S* represent the number of seats in that row. Can you give an equation for finding the number of seats?
d) How many seats are in Row 21? If the last row has 100 seats, how many rows are in the concert hall?

Paper Folding Problem (exponential)
Fold this piece of paper in half, and then open it up. How many regions were made?
a) Fold the paper in half twice. How many regions? How many regions will be made with three such folds?
b) Describe how to find the number of regions for any number of such folds.
c) Write an equation for finding the number of regions if you know the number of folds. Let *R* represent the number of regions, and let *F* represent the number of folds.
d) Suppose we have a magical piece of paper that can be folded indefinitely. If you fold the paper in half 10 times, how many regions are formed?

Car Wash Problem (inverse variation)
Metro Car Sales hires washers to wash the cars in their lot. The manager knows that 36 washers can wash all the cars in 1 hour.
a) What if only half as many washers worked? How long would it take 18 washers to wash all the cars? How long for 72 washers? How long for 9 washers? How long for 4 washers?
b) Describe how the manager could figure out how many hours it would take to wash the cars for any size group of washers.
c) Write an equation for finding the number of hours it takes to wash the cars if you know the number of washers. Let *H* represent the number of hours it takes, and let *S* represent the number of washers.
d) Can you use your equation to find how long it would take 10 washers?

Figure 24.1. Interview items that require (a) computing specific cases, (b) describing relations, (c) representing symbolically, and (d) using equations.

Grade 6 Students' Preinstructional Use of Equations to Describe and Represent Problem Situations   199

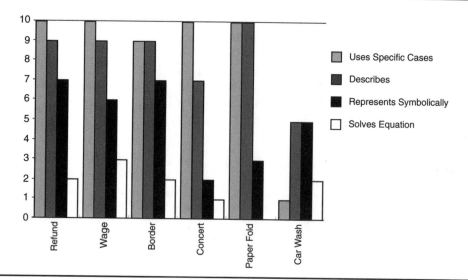

Figure 24.2. Interview items that require (a) computing specific cases, (b) describing relations, (c) representing symbolically, and (d) using equations.

problems. For example, for the Wage Problem a student said, "You two times how much overtime she has had, and then you add that to 20." For the Concert Hall Problem and the Paper-Folding Problem, a majority of the students described the relationships, but most did so recursively. For example, when the interviewer asked, "Can you describe how many regions there would be for any number of folds?" a student responded, "Uhm, it would be the number of regions times 2." The inverse-variation relation in the Car Wash Problem was the most difficult for students to describe. Only half the students were able to correctly describe the relationship, and they all described it in terms of the relationship between the independent and dependent variables.

Representing symbolically

All the students but one were able to generate an equation for at least one of the situations, although two students generated only one of the six equations. The student who was unable to represent any situation symbolically also had difficulty computing with specific numbers and describing the general relationships verbally. The most difficult situation for the students was the Concert Hall Problem, a linear situation represented by the expression $a(n-1) + b$. The relationships in the inverse-variation (Car Wash Problem) and the exponential (Paper-Folding Problem) situations also proved to be more difficult to generalize than those in the straightforward one- and two-step linear situations.

Using equations

Although at least half the students were able to consistently generate equations to represent the given situations, the students rarely seemed to use their equations as mathematical objects. However, two students explicitly used their equations to solve problems by reversing the operations represented in the equations. For example, for the Border Problem a student offered the equation ($n \times 4 - 4 = b$) to find the size of a grid, given the number of squares in the border.

DISCUSSION OF OUR FINDINGS

These sixth-grade students were able to numerically solve all the problems involving specific cases with the exception of the inverse-variation situation (Car Wash Problem). The sixth-grade students also demonstrated that they could describe relationships using verbal, symbolic, or a combination of verbal and symbolic representations. The difference in performance on these problems indicates that middle school instruction needs to push students' understandings beyond routine situations. Students seem able to generalize the arithmetic that they know well, but they seem to have difficulty generalizing the arithmetic with which they are less familiar. In particular, middle school students would benefit from more experiences with a rich variety of multiplicative situations, including proportionality, inverse variation, and exponentiation.

Lessons Learned From Research

There is potential benefit in examining the same problem through different representations, such as diagrams, graphs, tables, verbal descriptions, and equations. However, the *use* of multiple representations is not enough. For example, in the case of tables, the students could identify isolated patterns between pairs of dependent and independent variables but could not see a pattern that was consistent across the entire table. Tables were most useful when they were produced by students trying to make sense of the problem context instead of when provided or suggested by the interviewer. Whatever the representation, students need to make the link between the representation and the problem context and between one representation and another. At the prealgebra level, the emphasis in the curriculum should be on developing and linking multiple representations to generalize problem situations instead of on merely constructing representations that students do not link with problem situations. We need additional research to investigate the effect of innovative middle school curricula on enhancing students' abilities to use variables and generalize functional relationships.

REFERENCES

Edwards, E. L., Jr. (Ed.). (1990). *Algebra for everyone.* Reston, VA: National Council of Teachers of Mathematics.

National Council of Teachers of Mathematics. (1991). *Professional standards for teaching mathematics.* Reston, VA: Author.

CHAPTER 25

USING CONCEPT MAPS TO ASSESS CONCEPTUAL KNOWLEDGE

Carol G. Williams, Abilene Christian University

Abstract. I examined the value of concept maps as instruments for assessing conceptual understanding and for comparing the function knowledge of calculus students and that of mathematicians. Qualitative analysis of the maps reveals differences between the students and experts. Concept maps proved to be useful for assessing conceptual understanding. In this chapter I briefly describe what concept maps are and how they can be used to evaluate the extent to which an individual's knowledge of a concept, in this case functions, is connected.

USING concept maps is a direct method for looking at the organization and structure of an individual's knowledge within a particular domain and at the fluency and efficiency with which the knowledge can be used. The more connections that exist among facts, ideas, and procedures, the better one's understanding (Hiebert & Carpenter, 1992). Individuals whose knowledge within a particular domain is interconnected and structured will activate large chunks of information when performing an activity in that knowledge domain. A highly integrated knowledge structure signals the transition from novice to expert performance (Royer, Cisero, & Carlo 1993). In drawing and labeling the linking lines, the participants explicitly state the relationships they see. Mathematical knowledge and structure do not always lend themselves to simple categorizations, but they can be depicted well by concept maps.

This chapter is adapted from Williams, C. G. (1998). Using concept maps to assess conceptual knowledge of function. *Journal for Research in Mathematics Education, 29,* 414–421.

In my study, concept maps for function were drawn by volunteers enrolled in the third quarter of calculus. Each student attended an instructional session during which examples of concept maps were shown. Each map had concepts contained in ovals and linking words on the lines connecting the concepts. The examples included hierarchical maps, web or spider maps, and nonhierarchical maps. The students were instructed to draw their maps however they wished. Each student then drew up a list of terms related to functions and fashioned the terms into a concept map, adding other ideas when they arose. Each student completed the task in less than an hour.

Each of eight experts (PhDs in mathematics) drew up a list of starting terms before beginning the maps, just as the students had done, then drew a concept map of function to represent what they would *expect* students completing the first-year calculus sequence to know.

For this chapter I show two of the concept maps from my study. Figure 25.1 is a concept map drawn by a student from a traditional calculus class, and Figure 25.2 shows a concept map drawn by a mathematician. (Both have been redrawn here, but the nodes and connecting lines are just as in the originals.) I chose these concept maps to illustrate the manner in which they can illustrate an individual's knowledge connections for a particular mathematical topic.

INTERPRETING CONCEPT MAPS

The two maps illustrate the diversity and complexity of concept maps. Each map contains the function concepts in ovals with the words denoting

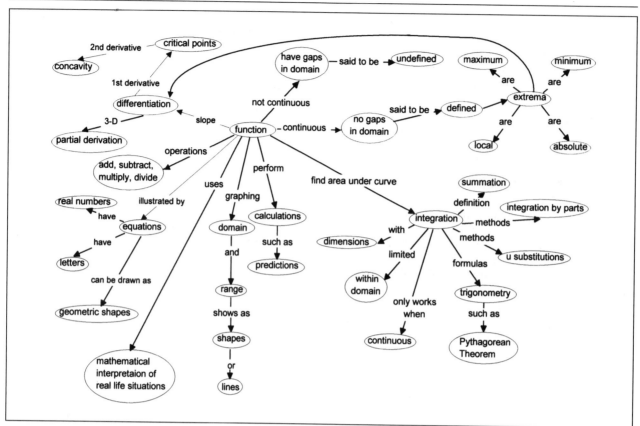

Figure 25.1. A traditional student's concept map of function (redrawn).

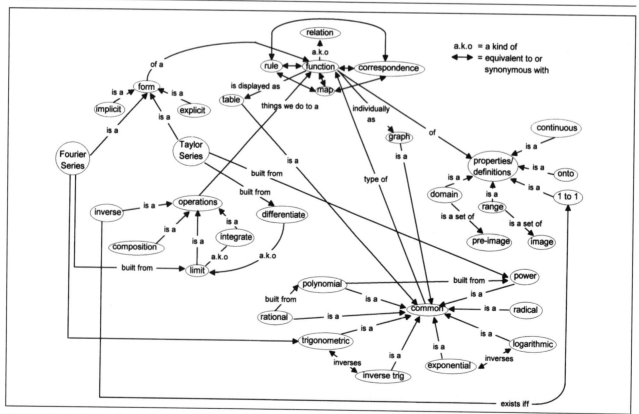

Figure 25.2. An expert's concept map of a function (redrawn).

Lessons Learned From Research

relationships among concepts on the lines linking the ovals. This common understanding of the process for drawing a concept map indicates that the wide diversity of maps derives mainly from different concepts of function.

I focus here on interpreting maps of functions by making four observations. First, a novice (such as the student whose map is shown in Figure 25.1) includes trivial or irrelevant concepts, for example, listing letters to indicate variables. Sometimes people who lack full understanding of a concept become preoccupied with the trees rather than with the forest. Second, students' maps are often algorithmic in nature; that is, a student might provide the steps for a procedure rather than name concepts and the relations that connect the concepts. But concept maps of experts are likely to reflect many categorical groupings rather than procedures. For instance, five of the experts had a grouping that referred to *classes* or *common types* of functions, using such terms as *exponential, polynomial, trigonometric, logarithmic,* and *rational.*

Third, the map maker's predominant view of the concept in question can be seen. For example, a student might say, straightforwardly, "Functions are *equations.*" But none of the experts demonstrated this propensity to think of a function as an equation. Instead, they defined *function* as *a correspondence, a mapping, a pairing,* or *a rule.* Fourth, the maps reflect the individuals' knowledge about the definition of the concept in question. For example, only a few students indicated that each element of the domain can be paired with only one range element. But all experts incorporated a definition of function into their maps. In overall content and complexity, the experts' maps as a group showed much more homogeneity than the students' maps.

The degree to which concept maps describe a person's actual mental representation is, of course, impossible to know. Nevertheless, the general homogeneity of the experts' maps and their distinct variance from the students' maps lend credibility to the conclusion that concept maps do capture a representative sample of conceptual knowledge and can differentiate well among fairly disparate levels of understanding. Information about students' understanding that an analysis of the maps provided is not readily gained from traditional pencil-and-paper tests. Concept maps, therefore, can provide important information about conceptual understanding, and I recommend their use in mathemat-ics classes to test knowledge of other concepts, such as rational numbers or two-dimensional geometric figures.

REFERENCES

Hiebert, J., & Carpenter, T. P. (1992). Learning and teaching with understanding. In D. A. Grouws (Ed.), *Handbook of research on mathematics teaching and learning* (pp. 65–97). New York: Macmillan.

Royer, J. M., Cisero, C. A., & Carlo, M. S. (1993). Techniques and procedures for assessing cognitive skills. *Review of Educational Research, 63,* 201–243.

CHAPTER 26

STUDENTS' CONCEPTUAL KNOWLEDGE OF FUNCTIONS

Brian R. O'Callaghan, Southeastern Louisiana University

Abstract. The foundation for this research on effects of *Computer-Intensive Algebra* (CIA) and traditional algebra curricula on students' understanding of the function concept is a proposed conceptual framework to describe function knowledge in terms of 4 component competencies: modeling, interpreting, translating, and reifying. The results indicated that the CIA students achieved better overall understanding of functions and were better at these 4 competencies than students from a traditional curriculum. Further, the CIA students showed significant improvements in their attitudes toward mathematics, were less anxious about mathematics, and rated their class as more interesting. In this brief chapter, the function components are described and some tasks used to assess each component are presented. I refer you to the original article for the description of the study itself.

FIRMLY rooted in a problem-solving environment, the function framework consists of four component competencies: modeling, interpreting, translating, and reifying. These competencies are accompanied by an associated set of procedural skills, all of which are described here.

Modeling

The process of mathematical problem solving involves a transition from a problem situation to a mathematical representation of that situation. This process entails using variables and functions to form an abstract representation of the quantitative

This chapter is adapted from O'Callaghan, B. R. (1998). Computer-Intensive Algebra and students' conceptual knowledge of functions. *Journal for Research in Mathematics Education, 29,* 21–40.

relationships in that situation. I refer to the ability to use functions to represent a problem situation, the first component of understanding of the function concept, as *modeling*. This component can be divided into a number of subcomponents depending on the representation system used to model the situation. The three most frequently used representations for functions are equations, tables, and graphs.

Interpreting

The reverse procedure is the interpretation of functions, in their different representations, in terms of real-life applications. This ability, labeled *interpreting*, is the second component of the model. This component also can be analyzed at a finer grain size and partitioned into subcomponents, which would again correspond to each of the three main representation systems for functions. In problems, students could be required to make different types of interpretations or to focus on different aspects of a graph—for example, on individual points versus more global features.

Translating

As mentioned previously, the mathematical model can be represented in various representation systems. The most commonly used forms are symbols, tables, and graphs, which have been called the three core representation systems. The ability to move from one representation of a function to another, or to *translate*, is the third component in the function framework.

Reifying

The final component of the framework for functions is *reification*, defined as the creation of a mental object from what was initially perceived as a process or procedure. For example, if a student thinks of a function as a process, she might think that $y = 3x + 4$ indicates a procedure for finding corresponding values for x and y. But if the student comes to understand function as an object, she can understand how such processes as transforming functions or finding the composition of functions can be applied *to* the function. That is, she now perceives the function as an object upon which one can operate.

Procedural skills

Associated with these four components of conceptual knowledge of functions is a set of proce-

dural skills. These skills consist of transformations and procedures that students can use to operate within a mathematical representation system. This set of skills is the primary area of concentration in the traditional curriculum, and it represents a large part of what algebra is to many students and teachers. But divorced from a conceptual understanding of functions, the procedural skills are not useful.

In the accompanying table (Table 26.1), I provide examples of tasks that can be used to analyze student understanding of each of the four components of my function model.

Table 26.1
Sample Questions From the Functions Tests

Component	Question
Modeling	A truck is loaded with boxes, each of which weighs 20 pounds. If the empty truck weighs 4500 pounds, find the following; a. The total weight of the truck if the number of boxes is 75. b. The number of boxes if the total weight of the truck is 6,740 pounds. c. Using W for the total weight of the truck and x for the number of boxes, write a symbolic rule (or equation) that expresses the weight as a function of the number of boxes.
Interpreting	The graph below (see Figure 26.1) gives the speed of a cyclist on his daily training ride. During his ride, he must climb a hill. He pauses for a drink of water before descending. Use this graph to answer the following questions as accurately as possible. a. Find the speed when time equals 25 minutes. b. Find the time when speed equals 30 mph. c. During what time intervals was the speed increasing? d. During which 10-minute interval did the speed decrease the most?

Figure 26.1. Graph of cyclist's speed versus time.

	e. When was the cyclist at the top of the hill?
Translating	Suppose that the following table gives the value (V), in dollars, of a car for different numbers of years (t) after it is purchased. Write a symbolic rule expressing V as a function of t.

t	V
0	$16,800
2	$13,600
4	$10,400
6	$7,200
10	?

Translating	A roast is taken from the refrigerator and put into an oven. The following is a table of its temperature, in degrees (D), recorded at different times during the first 120 minutes (m).

m	0	10	20	30	60	120
D	50	100	140	170	200	220

Draw a graph to represent this situation.

Reifying	A small company determines its contribution to charity (C) by its profit (p), which is dependent on the number of items (n) sold according to the following formulas:

$$C = .10(p - 1000) \quad \text{and} \quad p = 100n - n^2$$

a. What will the company contribute to a charity that sells 50 items?
b. Write a formula expressing C as a function of n.

CHAPTER 27

USING A CARD SORT TO DETERMINE ONE'S UNDERSTANDING OF FUNCTION

Gwendolyn M. Lloyd and Melvin (Skip) Wilson
Virginia Polytechnic Institute and State University

Abstract. An experienced high school mathematics teacher communicated deep and integrated conceptions of functions, dominated by graphical representations and covariation notions, in his first implementation of reform-oriented curricular materials during a 6-week unit on functions. These themes played important roles in the teacher's practice when he emphasized the use of multiple representations to help students understand dependence patterns in data. The teacher's well-articulated ideas about features of a variety of relationships in different representations supported meaningful discussions with students during the implementation of an unfamiliar classroom approach to functions. One way the teacher's notions about functions were determined was through a card sort.

IN OUR study of the influence of one veteran high school teacher's (Mr. Allen) conceptions of mathematical content on his teaching of a 6-week functions unit using curricular materials of the Core-Plus Mathematics Project, we addressed two questions: What were Mr. Allen's conceptions of functions prior to teaching with the Core-Plus materials? How did his conceptions of the subject matter shape his first implementation of the Core-Plus unit?

One way that we assessed Mr. Allen's conception of function was to ask him to participate in a "function sort." He was given the task of interpreting and organizing a stack of 32 cards containing mathematical relationships that differed by family (e.g., linear, polynomial, trigonometric, etc.), representation (equation, graph, table, or verbal description), and various characteristics (e.g., functionality, continuity, surjectivity, etc.). The cards, along with the teacher's descriptions of them, are displayed in Figure 27.1. (The word problems themselves were on the cards, even though the problems are not displayed in Figure 27.1. See Figure 27.2 for examples of word problems used.) Like the other representations, the word problems represented different families and types of functions; for example, the following problem was labeled *quadratic* by the teacher:

Fred is deciding which size pizza is the best buy. He wonders how the area of the pizza is related to its diameter.

Another problem, which the teacher categorized as *exponentially growing,* read as follows:

The 1990 census shows that Central City has a population of 40,000 people. Social scientists predict that Central City will experience a growth rate of 2% per year over the next 20 years. How can one predict Central City's population for each of the next 15 years?

This chapter is adapted from Lloyd, G. M., & Wilson, M. (1998). Supporting innovation: The impact of a teacher's conceptions of functions on his implementation of a reform curriculum. *Journal for Research in Mathematics Education, 29,* 248–274.

This study was supported in part by National Science Foundation Grant MDR-9255257. The opinions expressed are those of the authors and are not necessarily those of the Foundation.

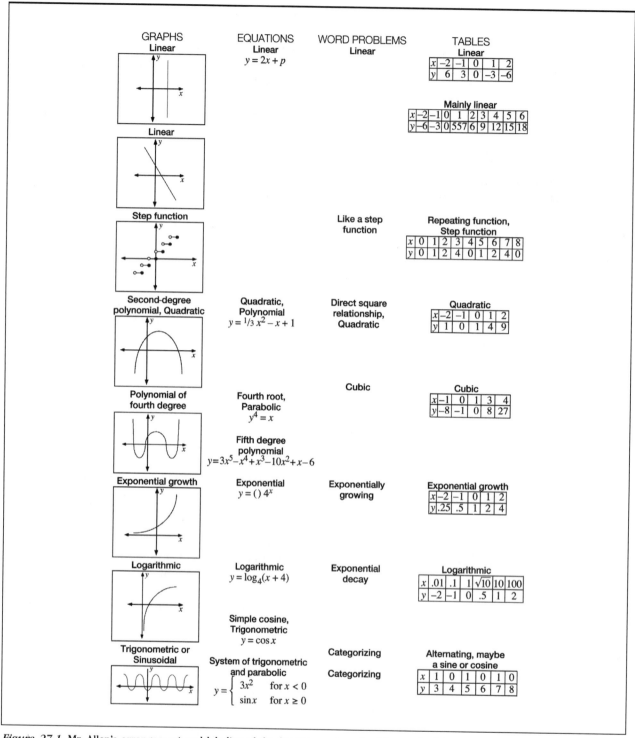

Figure 27.1. Mr. Allen's arrangement and labeling of the function-sort cards. (The cards did not contain the labels shown in the figure.) Within each representation group, he attempted to establish a similar ordering by family of relationship.

As these examples and those shown in Figure 27.1 reflect, the cards of the function sort incorporated diverse features and presented a challenging set of situations for the teacher to analyze and organize, as well as multiple criteria on which he could base different arrangements of the cards.

Given the task of interpreting and organizing the 32 cards of the activity, Mr. Allen first separated the cards into four piles according to the displayed representations, which he verbally named "graphs, word problems, equations, and tables," and then announced his decision to look more carefully at the

Lessons Learned From Research

Linear. Mr. Washington has noticed an increase of 3 cents per gallon in the price of regular unleaded gasoline over the past four weeks. If the current price is $1.309, he wonders what the price will be in the coming weeks if this same price increase continues each week.

Direct Square Relationship; Quadratic. Fred is deciding which size pizza is the best buy. He wonders how the area of the pizza is related to its diameter.

Exponential Growth. The 1990 census shows that Central City has a population of 40,000 people. Social scientists predict Central City will experience a growth rate of 2% per year over the 20 years. How can one predict Central City's population for each of the next 15 years?

Figure 27.2. A sample of the word problems used.

individual cards in the four representation groups to make sense of each relationship. His primary goal in looking at each relationship was to "classify the overall dependence pattern" shown on the card by answering the covariation question "As one variable changes, what happens to the other one?"

During his efforts to label the relationships, Mr. Allen demonstrated a strong preference for working with graphs and, to a somewhat lesser extent, equations. After beginning his labeling process with the group of equation cards, Mr. Allen picked up cards showing tables but then rejected them in favor of the graphs pile, explaining that he is "more of a visual person." He also demonstrated greater proficiency with graphical representations, as evidenced in the nature of his explorations of the variety of situations on the cards. On his first examination of the cards showing graphs and equations, Mr. Allen was quickly able to summarize the types. After looking briefly at the equation on each card, Mr. Allen was instantly able to summarize the types of relationships that he had identified: "It looks like I've got some linears and quadratic and exponential, a logarithmic and a root situation there, and some trig functions."

In contrast, his labeling of the word-problem and table cards required significantly more time and effort because he found the information on these cards to be less accessible. Mr. Allen made repeated assertions that his ability to identify the types of situations with these representations was simply "not as good" as with the graphs and equations. Graphs provided Mr. Allen with information that was difficult to retrieve from studying the values in tables, the representation with which he struggled most. For example, his observation of the

x- and y-values in a table led him to recognize that the x-values are "increasing by one" and "some square numbers [are] on the bottom" but did not enable him to satisfactorily answer his covariation question, "As x changes by one, what happens to y?" To help him "see" the relationship, Mr. Allen used the tabular values to draw a graph, then announced, "It's parabolic, centered around negative one," and concluded that the table represented a quadratic relationship.

Connections in Mr. Allen's organization of types of relationships

During the function sort, Mr. Allen created a two-dimensional organization of the cards that revealed meaningful conceptions not only of a variety of families and representations of functions but also of the connections among them. His identification of similar types of relationships in the four representation groups encouraged him to "group the cards in terms of some general classifications." He attempted to build these general classifications into an ordering that would provide a lateral infrastructure to accentuate the connections among groups of cards. Mr. Allen's final arrangement of the cards, with the descriptive labels that he created, is depicted in Figure 27.1.

The basic framework of Mr. Allen's ordering was shaped by his experience with the traditional secondary school curriculum. In every representation group, Mr. Allen identified those cards that describe the elementary families of function found in the traditional curriculum: linear, polynomial, exponential and logarithmic, and trigonometric. These families, in this order, compose his core arrangement of the cards in all four representation groups. This order corresponds to the sequence of Mr. Allen's teaching and learning of these topics in the traditional curriculum, which posits the linear model as "the most basic one that's taught."

When more than one card belonging to a particular family of functions appeared in a representation group, Mr. Allen invoked a sublevel to his main organization and ordered those cards by "simplicity." In contrast with his ordering of the core families, this ordering did not necessarily correspond to a teaching sequence. For example, Mr. Allen positioned his two linear graphs in the order shown in Figure 27.1 because he viewed the vertical line graph as showing "less change." However, Mr. Allen said that he would not typically introduce a vertical line graph to students first but would instead present it as a "special case" of the general linear model that the second graph exemplifies.

When Mr. Allen encountered a card situation that did not appear to correspond suitably to one of the elementary families of functions, he positioned it immediately following the most similar core family in his ordering. For example, Mr. Allen positioned his "mainly linear" table, for which he produced a graph that was linear with one exceptional point, immediately following his linear table. In this and subsequent cases, Mr. Allen used "the way the graph looks" to find the most appropriate position for a card. In other words, to position the outstanding cards in the sort, including step functions and nonfunctions, Mr. Allen needed to translate them to graphical formats. By relying on the graphical representation to solve the problem of placing the cards with challenging features into his ordering, Mr. Allen once again invoked his visual facility.

Mr. Allen's final arrangement of the function-sort cards illustrated his discernment and application of important connections among varied representations of families of relationships. By identifying the core families in each representation group, he demonstrated his ability to work with different displays of the same type of relationship. Additionally, his similar ordering of the core families in the graph, equation, word-problem, and table groups drew attention to the links among representations and also communicated the ties that he perceived, on the basis of teaching sequence and simplicity, among the different families of relationships.

This card-sort interview helped us understand Mr. Allen's conceptions of functions and thus helped us interpret his instruction when we observed his teaching. The card-sort activity can provide an interviewer a great deal of information about another person's thinking about functions. The basic structure of this card sort can be extended to other mathematical topics and can be used to examine another person's understanding of those topics.

CHAPTER 28

CROSS-NATIONAL COMPARISON OF REPRESENTATIONAL COMPETENCE

Mary E. Brenner, Sally Herman, Hsiu-Zu Ho, and Jules M. Zimmer
University of California at Santa Barbara

Abstract. Flexible use of multiple representations has been described as an essential component of competent mathematical thinking and problem solving. In this study, 6th-grade American students are compared to 3 samples of Asian 6th graders (Chinese, Japanese, Taiwanese) to determine whether the well-documented mathematical achievement of students from these Asian nations may be due in part to their greater understanding of mathematical representations. The results showed that among all groups, Chinese students generally scored highest on the representation tasks and, except on questions about the visual representations of fractions, all Asian samples scored significantly higher than the U.S. sample. The results are discussed in terms of possible instructional antecedents and textbook differences.

THE well-documented national differences in mathematics achievement provide rich opportunities for exploring the competencies that underlie mathematical performance. Although some people who compare mathematical achievement emphasize the number of facts that students from different countries know, in most recent documents *mathematical competence* has been defined more broadly. For instance, in *Principles and Standards for School Mathematics* (*Principles and Standards*) (National Council of Teachers of Mathematics, 2000) the emphasis is that "learning mathematics with understanding is essential" (p. 20). From this viewpoint mathematical competence consists of integrated factual knowledge, procedural proficiency, and conceptual understanding. *Principles and Standards* further indicates that mathematical problem solving is one means by which students develop mathematical understanding.

A skill important for problem solving is the ability to use a variety of representations to represent a problem. *Principles and Standards* has identified *representations* as one of the five process standards that should be emphasized throughout mathematics curriculum. The ability to use multiple representations deepens students' mathematical understanding and enhances their problem-solving skill.

In this study we examined the problem-representation skills of students from the United States, China, Taiwan, and Japan. In prior studies, students from the three Asian nations had demonstrated superior mathematical achievement when compared to U.S. students. However, the ways in which mathematics is taught in these countries differ significantly, and students' mathematical competencies might differ as a consequence.

This chapter is adapted from Brenner, M. E., Herman, S., Ho, H. -Z., & Zimmer, J. M. (1999). Cross-national comparison of representational competence. *Journal for Research in Mathematics Education, 30,* 541–557.

We thank our collaborators in the participating countries: Sou-Yung Chiu, National Chan-Hwa University of Education, Republic of China; Yasuo Nakazawa, Seisen Women's Junior College, Japan; and Chang-Pei Wang, Beijing Institute of Education, People's Republic of China.

THE VALUE OF REPRESENTATIONAL SKILL

Studies of representational skills are consistent in showing that U.S. students, as well as Asian students, can competently use representations to solve problems. However, both Japanese and Chinese students chose to use what other authors have called *more advanced* symbolic forms of representations, whereas U.S. students seemed to prefer visual ones.

Flexibility in moving across representations is a hallmark of competent mathematical thinking. Each type of representation highlights specific aspects of a concept and can help students in problem solving. This flexibility can entail moving within a representational mode, such as using a diagram with different arrangements of components to represent different proportion problems. It also entails moving between quite different representations, such as between an equation and a graph. Each representation reveals a different structural aspect of the problem. Solving difficult problems may require using several representations.

For this study, three aspects of representation use were examined. The first two involved transformations within the written-symbol-representation system. We made a distinction between Flexibility Within a Representation and Flexibility Across Formal Symbolisms (but still within a single representation system). As one example, we tested the former flexibility by asking students to express fractions as the combination of other fractions. This type of skill has been linked with mathematical competence in such areas as mental computation. The second type of flexibility, the ability to translate among different kinds of written representations, is believed to contribute to greater conceptual knowledge and enhanced problem solving. To test this ability, we asked students to, for example, express a fraction as a decimal, a division expression, and a proportional relationship. The third aspect that we examined was skill in translating between visual representations and written symbols, because earlier studies had shown that U.S. students were more likely than Asian students to use visual representations.

THE METHOD USED IN OUR STUDY

The participants in this study included 895 sixth-grade students from the four nations: 223 from the People's Republic of China, 224 from Taiwan, 177 from Japan, and 271 from the United States. The schools included in this study represented the range of schools in each country, including both high- and low-achieving schools and both urban and rural schools.

For this study we developed a mathematical-achievement test consisting of two types of questions (see Figure 28.1). Solution questions were either straightforward computations (S2, S3, and S5) or complex problems (S1 and S4). For each solution question, we wrote a set of representation questions (the two types of questions are paired in Figure 28.1). The students were asked to judge whether each representation question was a correct or incorrect representation for the corresponding solution question. (We did not include a solution question for the representation question 4×9 because this question was judged too easy for sixth graders.) The students were asked to complete the solution questions that appear in the first column of Figure 28.1 during one session and to complete the representation items appearing in the second column during a second session about 2 months later.

RESULTS OF THE STUDY

Solution Questions

Among the four national samples, the Chinese students ranked first on three of the five solution questions, and the Taiwanese students ranked second on four of the five. The Japanese sample ranked third on three of the five. The U.S. sample ranked last on all solution questions. Most Chinese students were able to accurately solve all but one solution question, and the Taiwanese students showed a similar rate of accuracy. In contrast, a slim majority of the U.S. sample answered two questions correctly with small percentages of the students answering the other three correctly. Thus the results of this study are similar to those of many others in that the Asian samples showed higher mathematical performance than the U.S. sample.

Representation Questions

Overall, students had more difficulty with representation questions than with solution questions. However, the pattern among samples was similar to the pattern for the solution questions: The Taiwanese and Chinese samples ranked first or second on most questions, with the Japanese sample showing the third-strongest performance overall. Students in the U.S. sample had strikingly low performances on the representation questions, scoring with less than 50% accuracy on more than 1/3 of the questions and ranking last among the nations on about 70% of the questions. In comparison, a major-

Solution Questions
(Student solves the problems)

S1 A pole 2 yards high casts a shadow 3 yards long. The shadow of a tree is 9 yards long. How tall is the tree?

S2 Give the answer to the following problem in decimals.

$$3 \div 6 = \underline{\hspace{1cm}}$$

S3 $1\frac{3}{4} + 2\frac{1}{2} =$

S4
In the rectangular diagram of a garden two white paths were made. What is the total area of the shaded sections below?
Answer _____

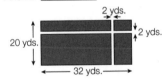

S5 $\frac{2}{3} \div \frac{2}{1}$

Corresponding Representational Items
(Student marks as right or wrong)

R1 (The same problem is stated.) "Now decide if each of the following statements is right or wrong. For each statement, check the appropriate box." (For each choice, boxes labeled Right and Wrong.)
a. $\frac{2}{3} = \frac{h}{9}$
b. $2:3 = h:9$
c. $3 + 2 = 9 + h$
d. $\frac{2}{9} = \frac{h}{3}$
e. $2 \times 9 = h \times 3$

R2 The fraction $\frac{3}{6}$ may be represented in several different ways. Decide if each of the following examples is a right or wrong representation of $\frac{3}{6}$.
a. $3 \div 6$
b. .50
c. ▪□▪▪□
d. ⬚ : ⬚
e. ◔
f. Paul has half a dozen doughnuts. He wants to share them equally with his friend Joani.
g. ▪▪□□□□□□□

R3 Given the problem below, decide if each of these statements is right or wrong.
$$1\frac{3}{4} + 2\frac{1}{2} =$$
a. $\left(1\frac{3}{4}\right) + \left(2\frac{1}{2}\right)$
b. $\frac{4}{4} + \frac{3}{4} + \frac{4}{2} + \frac{1}{2}$
c. $\left(1 \times \frac{3}{4}\right) + \left(2\frac{1}{2}\right)$
d. $1 + 2 + \frac{3}{4} + \frac{1}{2}$
e. $1 + 0.75 + 2 + 0.50$

R4 (The same figure is given.)
a. The total area of the figure is 20×32.
b. The area of the white paths is $(2 \times 20) + (2 \times 32)$.
c. The area of the shaded sections is $(20 \times 32) - (2 \times 20) - (2 \times 32)$.
d. The area of the shaded sections is $(20 \times 32) - (2 \times 20) - (2 \times 32) + (2 \times 2)$.
e. The area of the shaded sections is 18×30.

R5 Read the problem below.

Paul and his friend Mathew love to eat pizza. They are given two thirds of a whole pizza to share equally. What portion of the whole pizza would each person get to eat? Now decide if each of the following statements is right or wrong.

a. $\frac{2}{3} + \frac{1}{2}$
b. $2 \div \frac{2}{3}$
c. $\frac{2}{3} \div 2$
d. $\frac{2}{3} \times \frac{1}{2}$
e. $\frac{2}{3} \div \frac{1}{2}$

(continued on next page)

Figure 28.1. Test questions.

Cross-National Comparison of Representational Competence

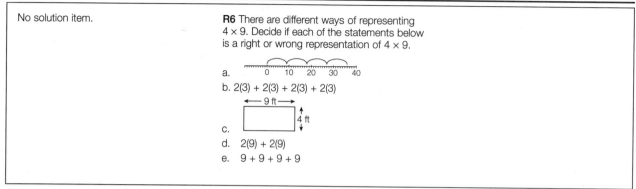

No solution item.

R6 There are different ways of representing 4 × 9. Decide if each of the statements below is a right or wrong representation of 4 × 9.

a.

b. 2(3) + 2(3) + 2(3) + 2(3)

c.

d. 2(9) + 2(9)

e. 9 + 9 + 9 + 9

Figure 28.1 (continued). Test questions.

ity of Chinese and Taiwanese students correctly answered the representation questions on all but one set of questions.

Flexibility within a representation

Nine questions required students to judge the equivalence of different fraction representations: R3a–d and R5a–e. The Chinese sample consistently scored highest, and the U. S. students scored lowest. Fewer than half of the U.S students correctly answered most of these questions, thus producing the lowest average performance of the national samples.

Flexibility across formal symbolisms

The questions related to this type of flexibility were mixed and included (a) transformations of fraction to decimal, fraction to proportion, and multiplication to addition and (b) translations of multiplication to a number line and of simple number sentences to algebraic number sentences. (For this analysis, a question was judged to use algebraic notation when it involved a variable or used parentheses for grouping different operations.) The 12 questions analyzed for this kind of flexibility were R1a–d; R2b and d; R4c and d; and R6a, b, and d. On average these were the most difficult representation tasks. The Chinese and Taiwanese samples attained the highest scores on these difficult questions, although even their performances were often little better than would be achieved by guessing. Once again the U.S. sample had the lowest scores.

In a number of cases, despite showing computational skill with questions in various forms, the students seemed to reject the very idea that these forms of formal notation could be equivalent. For instance, except for the U.S. sample, all national samples did very well calculating a decimal answer for 3 ÷ 6 (87% to 97% accuracy on S2). However, they rejected the idea that .50 was an appropriate representation for 3/6 on (in R2b), although they saw the equivalence between the fraction and the division computation (in R2a). The use of algebraic notation in many of these questions may have contributed to the students' difficulties. The three most difficult representation questions in the study (R1e, R4d, R6d), on which students had less than 50% accuracy overall, used algebraic notation. This pattern of difficulty held for all four national samples.

Visual representations

Two solution questions (S1 and S4) included diagrams that could facilitate solving the problems. Despite U.S. students' preference for visual representations in earlier studies, in this study their scores on these questions were significantly lower than the scores of the children from the other three nations. In fact, U.S. students' scores on these two questions were their lowest among all solution questions; they scored only 18% and 10% correct on S1 and S4, respectively.

Five of the representation questions (R6c, R2c, R2d, R2e, and R2g) were visual, that is, diagrammatic, representations of the solution questions. On Questions R2c and R2g, each of which was a rectangle with regions shaded to represent a fraction, the U.S. sample scored .70 and .69, respectively, significantly higher than the samples of the other nations. However, on two other visual questions, the U.S. sample had the lowest accuracy score among the four samples.

Lessons Learned From Research